Preface

As the world stumbles on in the new century, the doubts about man made climate change are clearly being dispelled by hard experiences of floods, droughts, wild fires and environmental degradation. The world climate is spiralling into a radical new phase not known since before records begun. The resulting deprivation of whole populations are causing bickering and hostilities between neighbouring communities, and a world wide instability that is fuelling south – north migration and much animosity and xenophobia.

We are in the final chance, no excuses, no prevarication, fossil fuels have to be curbed, then halted, land development must be controlled with respect to the natural environment and the exploitation of natural resources and the high seas and inland mining need international legislation, regulation and control.

The last 50 years have seen massive strides in electronics, computing and communications, such that on a personal level an average citizen has access to a smart phone with all the computing and data handling power of what previously would be considered national computing centres. The result is that people have access to communication worldwide and may now instantly experience what is happening on the other side of the world leading to many people in developing nations feeling the pressure of being neglected, disenfranchised, disadvantaged underprivileged and in turn are willing to risk life and limb to improve their situation.

Naturally, privileged people want to keep their privileges and will justify any action that supports this. In developed nations many are choosing the radical nationalistic right and there is a polarisation of opinion into no compromise positions between those that prioritise markets and the economy at any cost including disregard to the environment and poverty and they would war on immigration and the so called 'liberals ' who insist on due consideration of these issues and point to the root causes and the needs for social intervention.

For the last 40 years the economists mantra of 'growth and more growth' has led to climate change. This growth has been fed by coal consumption for production of cement steel and bricks, and consumption of oil and gas for transportation of goods worldwide personal transportation, household heating and industrial expansion and a total disregard for adherence to environmental controls. It is clear any further economic growth cannot be made at the expense of more fossil fuel consumption.

Now that everyone has their own computer in their fingers (even though it may only be their smartphone or tablet) there is the opportunity to avail oneself of the huge amount of data, ideas and processing power on the Internet and use what one finds as a guide for information. There are many proprietary and commercial packages available for tasks that work seamlessly towards their often specialist functions. The issues with such packages are they are easily directed and lead the user to

1

a very blinkered and specific use, not allowing diversions and developments from alternative perspectives.

There are also a whole bulk of materials that come free or inexpensive as freeware, they are made available either for promotional reasons or for reasons of education or for philosophical reasons of dissemination. Examples of these may be found on sites such as Sourceforge (https://sourceforge.net) and CNET (https://download.cnet.com), prominent amongst these are the Gnu/Linux systems and Opensource, these contain excellent tools that you are encouraged to use, adapt and develop to your specific requirement. Some of these tools I use in this project, to demonstrate that with the power of a modest modern machine and a low budget you can still derive a lot of pleasure developing potent applications and get remarkable results.

In this document I take a look at three interesting topics and show how our own home computer / laptop is a powerful machine.

Synopsis

I begin by examining the mathematical ideas of series. How do we obtain the sums of arithmetic and geometric series . We set out basic theory of series and convergence. We use this to show the harmonic series is divergent and we use this approach to to test the convergence of other examples.

I show how the area of a 30° arc can be used to generate a Newtonian expansion for accurately calculating pi. Such, progresses term by term and rapidly converges. This is compared to a similar algorithm produced by measurement angle in radians and using the asin function resulting in a similar solution. I use my system of extended precision to calculate these values and demonstrate the exponential relation between time and precision of calculation. Calculating pi using the Machin formula still proves to be efficient. I then introduce the GMP system for arbitrary precision calculation. I create a suite of routines that can replace the operations of my extended precision system. Using the Machin formula in the GMP system improves speed several thousand fold.

The use of Newton expansion shows we have a tool for accurately obtaining integrals of functions related to surds and expressions that are easily expanded into power series.

I make a study of regular polyhedra which we generate from primitive regular polygons with sides of unit length. It is shown how to find the angle between the faces and from there how to create the complete polyhedra and how to centre it on the origin. We are introduced into the criteria that limit the possibilities for creating regular polyhedra and also the number available. I create models of these polyhedra and some of their derivatives that may be explored with 3d design software.

Radix Mission

Olatunde Adeyemo

2022

First Printing: 2022

www.anglo-african.com

Available from http://www.lulu.com

ISBN

978-1-4710-2353-8
Imprint: Lulu.com

Table of Contents

Illustrations

Series

Let us look at series. The simplest series is the repetition of 1 the sum of which is written as

$$\sum_{i=1}^{N} 1 = N$$

there is nothing much to say of this.

What of the sum of an arithmetic series? Consider the sequence of natural numbers.

$$\sum_{i=1}^{N} i = N(N+1)/2 \qquad\qquad 1.1$$

The sum of series result above is demonstrated below. By writing the series forward and in parallel writing the series in reverse and summing each combination it is easily shown the total is N*(N+1)

arithmetic Progression	1	2	3	4	5	6	7	8	9	10
reverse Order	10	9	8	7	6	5	4	3	2	1
sum	11	11	11	11	11	11	11	11	11	11

total = 10*11

The sum of i^2.

$$\sum_{i=1}^{N} i^2 = \frac{N^3}{3} + \frac{N^2}{2} + \frac{N}{6} \qquad\qquad 1.2$$

We can obtain the proof of this. Let us look at $\sum_{i=1}^{N} i^3$

now $\sum_{i=1}^{N} (i+1)^3 = \sum_{i=1}^{N} i^3 - 1 + (N+1)^3$

but from binomial theory

$\sum_{i=1}^{N} (i+1)^3 = \sum_{i=1}^{N} i^3 + \sum_{i=1}^{N} 3i^2 + 3\sum_{i=1}^{N} i + N$ combining the two we can get

$3\sum_{i=1}^{N} i^2 = -N - 1 + (N+1)^3 - 3\sum_{i=1}^{N} i$

we know $(N+1)^3 = N^3 + 3N^2 + 3N + 1$

$$3\sum_{i=1}^{N} i^2 = N^3 + 3N^2 + 2N - \frac{3N(N+1)}{2}$$

$$\sum_{i=1}^{N} i^2 = \frac{N^3}{3} + \frac{N^2}{2} + \frac{N}{6} \qquad \text{QED.}$$

The same method may be used to show

$$\sum_{i=1}^{N} i^3 = \frac{N^4}{4} + \frac{N^3}{2} + \frac{N^2}{4} \quad 1.3$$

and so on, each next level relies on the revelation of the previous.

For a geometric progression.

$$\sum_{i=0}^{N} x^i = \frac{x^{N+1} - 1}{x - 1} \qquad 1.4$$

To show this

$$\sum_{i=0}^{N} 2^i = 1 + 2 + 4 + 8 + 16 + 32 + 64 \ldots 2^N$$

eg Below we show the progressive sums for various values of x.

i		1	2	3	4	5	6
2^i	1	2	4	8	16	32	64
sum	1	3	7	15	31	63	127
3^i	1	3	9	27	81	243	729
sum	1	4	13	40	121	364	1093
4^i	1	4	16	64	256	1024	4096
sum	1	5	21	85	341	1365	5461
5^i	1	5	25	125	625	3125	15625
sum	1	6	31	156	781	3906	19531
$(5^{(N+1)}-1)/(5-1)$		6	31	156	781	3906	19531

by examination we see the pattern of results to give eqn 1.4

If we have a decreasing series where for a given value N, n>N then for all terms $a_{n+1} < a_n$ .We may wish to know if the series converges to a sum as an important property. Thus series a_n is said to be convergent if

$$\sum_{i=1}^{\infty} a_i = S$$ the nth partial sum is then defined $$\sum_{i=1}^{n} a_i = S_n$$ and we can state $$\lim_{n \to \infty} S_n = S$$

We can now state a number of theorems that help us work work with convergent series.

Theorem 1.1
if
$$\sum_{i=1}^{\infty} a_i = S_a \text{ and } \sum_{i=1}^{\infty} b_i = S_b$$

then
$$\sum_{i=1}^{\infty} a_i + b_i = S_a + S_b \quad \text{and} \quad \sum_{i=1}^{\infty} a_i - b_i = S_a - S_b$$

ie if we have two convergent series then their sum and difference are also convergent.

Theorem 1.2

the series $$\sum_{i=1}^{n} a_i = S_n$$ is divergent if $$\lim_{n \to \infty} a_n \neq 0$$, then S is unbounded.

Theorem 1.3

If $$\sum_{i=1}^{\infty} b_i$$ is a convergent monatonic decreasing series and there is a positive integer N such that

n >N, series $a_n <= b_n$ then $$\sum_{i=1}^{\infty} a_i$$ is also convergent.

Conversely, if $$\sum_{i=1}^{\infty} b_i$$ is divergent monatonic decreasing series and there is a positive integer N such

that n >N, series $a_n >= b_n$ then $$\sum_{i=1}^{\infty} a_i$$ is divergent.

The chart below shows the relationship between the terms of a monatonic decreasing series and its associated real function. It clearly shows the area under the function is less than the summation of the terms $$\sum_{i=1}^{\infty} a_i$$, however this same area is greater than the summation of following terms $$\sum_{i=1}^{\infty} a_{i+1}.$$

From theorem 1.3 if the integral for the function can be obtained and is finite then the series $\sum\limits_{i=1}^{\infty} a_{i+1}$

converges and so also $\sum\limits_{i=1}^{\infty} a_i$ since there is only one term difference between the two.

Thus $\int\limits_1^{\infty} f(x)\,dx \geq \sum\limits_{i=2}^{\infty} a_i$ and $\sum\limits_{i=1}^{\infty} a_i \leq a_1 + \int\limits_2^{\infty} f(x)\,dx$.

Conversely if $\int\limits_1^{\infty} f(x)\,dx = \infty$ then $\sum\limits_{i=1}^{\infty} a_i$ is divergent.

This is known as the integration test.

The relation between f(x) , a_x and a_{x+1}

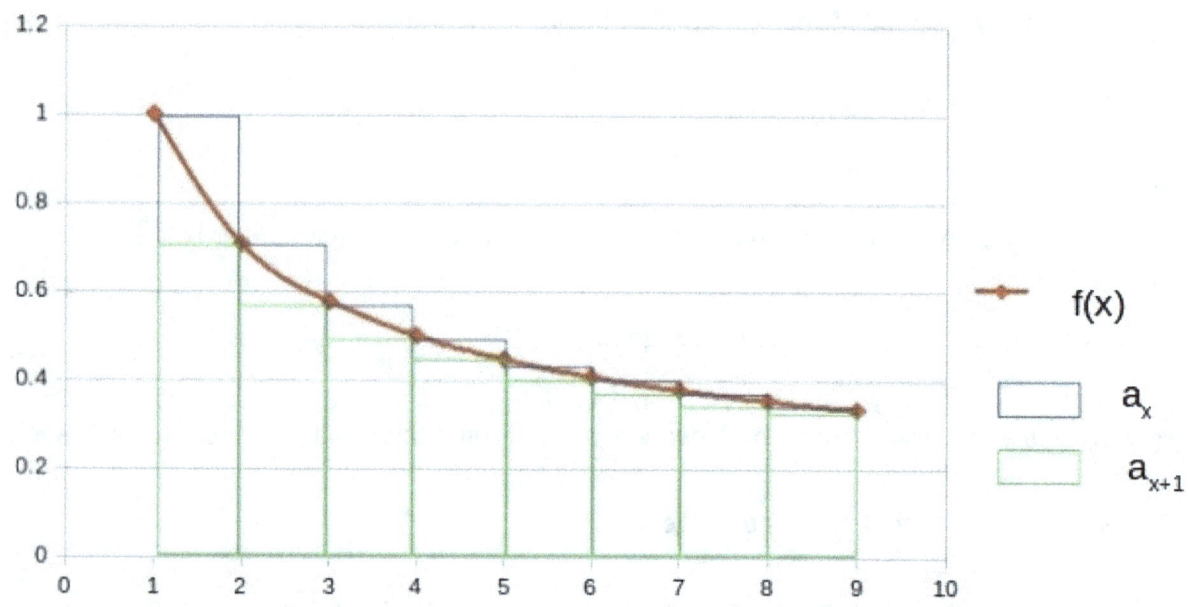

Theorem 1.4   The ratio test

If $\lim\limits_{n\to\infty} a_n \neq 0$ and $\lim\limits_{n\to\infty} \dfrac{a_{n+1}}{a_n} = L$ then if L >1, $\sum\limits_{i=1} a_i$ is divergent.

if L <1, $\sum\limits_{i=1} a_i$ is convergent.

If L=1 the test fails.

Theorem 1.5   The nth root test

If $\lim\limits_{n\to\infty} a_n \neq 0$ and there is a root such that $\lim\limits_{n\to\infty} \sqrt[n]{a_n} = L$ then

if L >1, $\sum\limits_{i=1} a_i$ is divergent.

if L <1, $\sum\limits_{i=1} a_i$ is convergent.

If L=1 the test fails.

If we consider the harmonic series.

$1 + \dfrac{1}{2} + \dfrac{1}{3} + \dfrac{1}{4} + \dfrac{1}{5} + \dfrac{1}{6} + \dfrac{1}{7}\ldots + \dfrac{1}{n}$ to answer if this series is convergent we may check against our theorems.

It is clear to see for this series $\sum\limits_{i=1}^{n} \dfrac{1}{i}$ and $\lim\limits_{n\to\infty} a_n = 0$ so theorem 1.1 is not true and from this theorem we cannot say the series is definitely divergent.

When take into account the the integration test we have the associated function is $\dfrac{1}{x}$ and we have

$\int\limits_{1}^{\infty} \dfrac{1}{x} dx = {}_1^\infty[\ln x] = \infty$ This concludes that the series is divergent.

This can be shown more demonstrably,

$$1 + \dfrac{1}{2} + \dfrac{1}{3} + \dfrac{1}{4} + \dfrac{1}{5} + \dfrac{1}{6} + \dfrac{1}{7}\ldots + \dfrac{1}{n} \ = \ 1 + \dfrac{1}{2} \ + \ \dfrac{1}{3} + \dfrac{1}{4} \ + \ \dfrac{1}{5} + \dfrac{1}{6} + \dfrac{1}{7} + \dfrac{1}{8} + \ldots + \ldots + \ldots + \ldots + \dfrac{1}{n}$$

$$\geq \ \dfrac{1}{2} + \dfrac{1}{2} + \dfrac{1}{2} + \dfrac{1}{2} + \dfrac{1}{2} + \ldots + \ldots + \ldots + \dfrac{1}{n}$$ which is clearly unbounded.

If we test with the ratio test we have $\lim\limits_{n\to\infty} \dfrac{a_{n+1}}{a_n} = \dfrac{n}{n+1} = 1$. This test is inconclusive.

Having shown $\sum\limits_{i=1}^{n} \dfrac{1}{i}$ is divergent, let us consider $\sum\limits_{i=1}^{n} \dfrac{1}{i^k}$. The integration test gives f(x) = $\dfrac{1}{x^k}$,

$\int\limits_{1}^{\infty} \dfrac{1}{x^k} dx = [\dfrac{(1-k)}{x^{(k-1)}}]_1^{\infty} = k-1$ if k>1 convergent, if k<1 the absolute value of the integral expands to

infinity and the series diverges.

Let us look at series of the form $\sum\limits_{i=1}^{n} \dfrac{1}{k^i}$. If k has an absolute value greater than 1 then $\lim\limits_{n\to\infty} \dfrac{1}{k^n} = 0$
and this does not confirm divergence of the series.

Let us look at the series in respect to integral test $\ln[\dfrac{1}{k^n}] = -n\ln k$ we can then test

$\int\limits_{1}^{\infty} x \, lnk \, dx = \dfrac{-\ln k}{2}[x^2]_1^{\infty} = -\infty$, as long as ln k >0 as before $\lim\limits_{n\to\infty} \dfrac{1}{k^n} = 0$ and the integral test shows

the series converges.

In respect to the ratio test, theorem 1.4 $\lim\limits_{n\to\infty} \dfrac{a_{n+1}}{a_n} = \dfrac{k^n}{k^{(n+1)}} = \dfrac{1}{k}$ Again as long as $|k|>1$ the test
shows the series converges.

In respect to the nth root test $\lim\limits_{n\to\infty} \sqrt[n]{a_n} = \sqrt[n]{\dfrac{1}{k^n}} = \dfrac{1}{k}$ Again as long as $|k|>1$ the test shows the
series converges.

Looking further into this series

i		1	2	3	4	...	n
$\dfrac{1}{k^i}$	$\dfrac{1}{k^1}$	$\dfrac{1}{k^2}$	$\dfrac{1}{k^3}$	$\dfrac{1}{k^4}$		...	$\dfrac{1}{k^n}$
sum	$\dfrac{1}{k^1}$	$\dfrac{(k+1)}{k^2}$	$\dfrac{((k+1)*k+1)}{k^3}$	$\dfrac{(((k+1)*k+1)*k+1)}{k^4}$		...	$\dfrac{\sum\limits_{i=1}^{n-1} k^i}{k^n}$

from eqn 1.4 $\sum\limits_{i=0}^{N} x^i = \dfrac{x^{N+1}-1}{x-1}$, making $\sum\limits_{i=1}^{N-1} \dfrac{1}{k^i} = \dfrac{k^N-1}{(k-1)k^N} - \dfrac{1}{k^N}$ and this results in :-

$$\lim_{n \to \infty} \sum_{i=1}^{n} \frac{1}{k^i} = \frac{1}{k-1}$$

Pi, but a lot faster.

Lets have a second look at calculating Pi. In "Alien Pi in the Sky" 2016 in chapter 3 we produce an extended precision arithmetical system that forms the basis of arbitrary precision calculations, we used it to calculate Pi using Gregory series in an efficient manner. In effect however we find it quite "cluncky" and slow.

I have since made several revisions of the programs used in the system, creating a basic core of routines and packaging them in a module that can be used to do maths calculations and input and output variables to the console or disk files. This module can be appended to a main program using the(#include "fnct.ext")instruction. In the main program file we still need to set out enough storage for each NxlongR type, the extended precision element, by setting the "#define ncells " statement and we declare the precision of our calculations by setting the value of Ndp.

This extended precision module with other programs can be found at my GIT repository Tunde-Adeyemo/arbitrary-p.

Using this new arrangement with a more modern 64 bit machine we may again run a program using the original method, we find there is little improvement in performance times. The points to be noted from this experience are that we have a system that can do general precision rational calculations to whatever precision we require and we have useful general functions such as cosin, sin, arctan, nth power root and we have the output to write our extended precision variables to file and input functions to read extended precision variables from a file.

Another approach to measuring Pi was made by Newton.
Consider the figure below. The 30° sector represents 1/12 area of the full circle, given as πr^2, since r =1, area A= $\pi/12$ and is calculated

$$A = \int_0^{0.5} \left(1-x^2\right)^{0.5} dx - 0.5*0.5*\left(3/4\right)^{0.5}$$

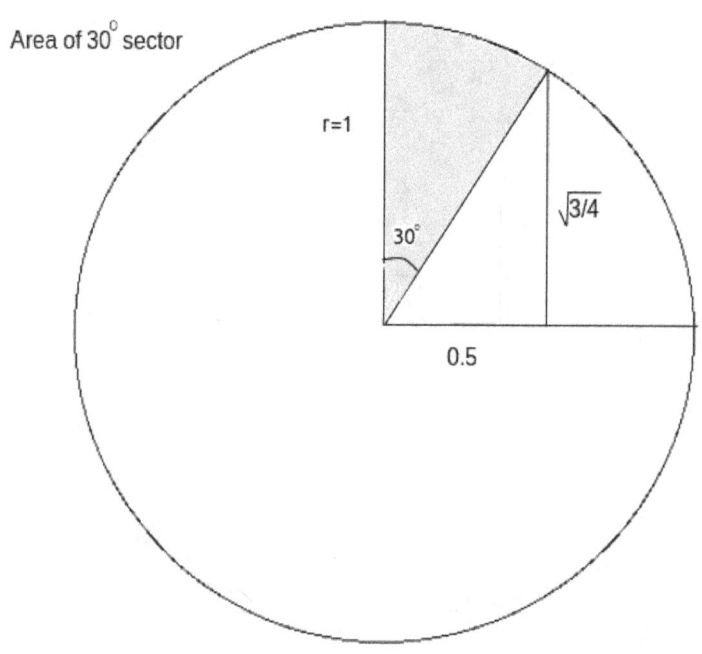

Area of 30° sector

r=1

$\sqrt{3/4}$

30°

0.5

to evaluate the integral, we may consider the binomial expansion

$$(a+b)^q = a^q + \binom{q}{1} a^{q-1} b + \binom{q}{2} a^{q-2} b^2 + \ldots \binom{q}{n} a^{q-n} b^n$$

where q is a rational and n is a positive integer.

$$\binom{q}{n} = \frac{q \cdot (q-1) \cdot (q-2) \ldots (q-n+1)}{n!} \quad 2.1$$

The trick is, no matter which rational value q takes, b is only raised to integer values. If we substitute 1 for a and $-x^2$ for b

$$(1-x^2)^q = 1 + \binom{q}{1} \cdot -x^2 + \binom{q}{2}(-x^2)^2 + \ldots \binom{q}{n}(-x^2)^n \quad 2.2 \text{ since each successive term varies by factor}$$

$$\frac{(q-n+1) \cdot -x^2}{n}, \quad \frac{(q-n+1) \cdot -x^2}{n} \lim n \to \infty = x^2$$

thus if -1< x<1 the terms will converge. We can then evaluate our integral by integrating term by term wrt x.

$$\int (1-x^2)^q \, dx = x - \binom{q}{1} \frac{x^3}{3} + \binom{q}{2} \frac{x^5}{5} - \ldots + \ldots \binom{q}{n} \frac{(-1)^n x^{2n+1}}{(2n+1)} \quad 2.3 \qquad \text{and A is then}$$

$$A = \int_0^{0.5} (1-x^2)^{0.5} \, dx = x - \binom{0.5}{1} \frac{x^3}{3} + \binom{0.5}{2} \frac{x^5}{5} - \ldots + \ldots \binom{0.5}{n} \frac{(-1)^n x^{2n+1}}{(2n+1)} - 0.5*0.5*(3/4)^{0.5} \quad 2.4$$

with each successive term converging if -1< x <1.

$$\binom{0.5}{1} = 0.5, \quad \binom{0.5}{2} = \frac{0.5*-0.5}{2}, \quad \binom{0.5}{n} = \frac{0.5*-0.5*-1.5 \ldots *(1.5-n)}{n!} \quad 2.5$$

$$\binom{0.5}{n} \text{ can be written as } \binom{0.5}{n} = \frac{1}{n!} \prod_{i=1}^{n} (1.5-n) \text{ and}$$

$$A = \pi/12 = \left[x - 0.5*0.5*(3/4)^{0.5} + \sum_{n=1}^{N} \left(\frac{-1^n * x^{2n+1}}{n!} \prod_{i=1}^{n} (1.5-n) \right) \right]_{x=0}^{x=0.5} \quad 2.6$$

So as long as we have an accurate value for $\dfrac{-\sqrt{3}}{8}$ we can get ever more precise values for A by adding successive terms. Thus we have produced a series solution for π that rapidly converges.

An alternative method is to use the direct measure of angle in radians thus for a complete rotation we subtend 2π radians and for $30°$ we subtend $\pi/6$ radians, $\sin 30° = 0.5$ so $\sin^{-1} 0.5 = \pi/6$.
Let us consider

$x = \sin \theta$ thus

$dx = \cos \theta \, d\theta = \sqrt{1-x^2} \, d\theta$

$\theta = \int_0^X \dfrac{dx}{\sqrt{1-x^2}}$ This is of the form of the binomial expansion given above.

$(1-u)^{-1/2} = 1 + \sum_{k=1}^{\infty} \binom{\frac{-1}{2}}{k} \cdot (-u)^k$

$\binom{\frac{1}{2}}{k} = \dfrac{-1/2. -3/2. -5/2 ... -(k+1)/2}{k!}$

with $u = x^2$ the above integral becomes

$\int_0^X \dfrac{dx}{\sqrt{1-x^2}} = \int dx + \sum_{k=1}^{\infty} \binom{\frac{-1}{2}}{k} \cdot \int (-x^2)^k \, dx$ 2.7

$\theta = x + \sum_{1}^{\infty} \binom{\frac{-1}{2}}{k} \cdot \dfrac{(-x^2)^k \cdot x}{2k+1}$ 2.8

Thus the error in our estimate will be of the order $\dfrac{x^{2k+1}}{2k+1}$ and we can work towards this.

Alternatively,

$\theta = \int_0^X \dfrac{dx}{\sqrt{1-x^2}}$

if $b^2 < u^2$

$f(u+b) \simeq f(u) + bf'(u) + b^2 \dfrac{f^2(u)}{2} + ... + b^i \dfrac{f^i(u)}{i!}$

as a Newton Raphson expansion

expand $(1-x^2)^{-0.5}$ about 1 when $-1 < x < 1$

16

$$(1-x^2)^{-0.5}=1-x^2(-0.5)+\frac{-0.5(-1.5)}{2}*(-x^2)^2....+\frac{-0.5(-1.5)...(-i+0.5)}{i!}*(-x^2)^i \quad 2.9$$

and

$$\int (1-x^2)^{-0.5}dx=x-\frac{-0.5\,x^3}{3}+\frac{-0.5(-1.5)(-x^2)^2x}{2*5}....+\frac{-0.5(-1.5)...(-i+0.5)(-x^2)^i x}{i!*(2i+1)} \quad 2.10$$

which may be written in more compact form

$$\int (1-x^2)^{-0.5}dx=x+\sum_{i=1}^{\infty}\frac{(-x^2)^i x\prod_{l=1}^{l=i}(0.5-l)}{i!(2i+1)} \quad 2.11$$

similarly we may expand

$$(1-x^2)^{+0.5}=1-x^2(0.5)+\frac{0.5(-0.5)}{2}*(-x^2)^2....+\frac{0.5(-0.5)...(-i+1.5)}{i!}*(-x^2)^i$$

and

$$\int (1-x^2)^{+0.5}dx=x-\frac{0.5(x^3)}{3}+\frac{0.5(-0.5)(-x^2)^2.x}{2*5}....+\frac{0.5(-0.5)...(-i+1.5)(-x^2)^i.x}{i!*(2i+1)}$$

which can be written more compactly.

$$\int (1-x^2)^{+0.5}dx=x+\sum_{i=1}^{\infty}\frac{(-x^2)^i x\prod_{l=1}^{l=i}(1.5-l)}{i!(2i+1)} \quad 2.12$$

The Newton-Raphson or binomial expansions given above can easily be adapted to functions of $1+x^2$.

From 2.2

$$(1-x^2)^q=1+\binom{q}{1}.-x^2+\binom{q}{2}(-x^2)^2+....\binom{q}{n}(-x^2)^n \quad 2.2$$

substitute +y for -x² to obtain

$$(1+y)^q=1+\binom{q}{1}.y+\binom{q}{2}y^2+....\binom{q}{n}y^n \quad 2.13$$

This leads to solutions to expansions for $\tan^{-1}$ in terms of $\tan\theta$, $x=\tan\theta$, $\sec^2\theta=1+\tan^2\theta$

17

and relations between sinh x and cosh x through the relation $\cosh^2 x = 1 + \sinh^2 x$.

let $u = \tan\theta$,

$$du = (\sec^2\theta)\,d\theta = (1 + u^2)\,d\theta$$

$$\int d\theta = \int \frac{du}{1 + u^2} \quad 2.14$$

as above, if $-1 < u < 1$ this is easily expanded to

$$\theta = \int (1 + u^2)^{-1}\,du = u + \sum_{i=1}^{\infty} \frac{u^{2i+1}\prod_{l=1}^{l=i} l}{i!(2i+1)} \quad 2.15$$

let u = sinh x,

du = cosh x dx =

$$\int dx = \int \frac{du}{(1 + u^2)^{0.5}} \quad 2.16$$

from 1.13

$$x = \int (1 + u^2)^{-0.5}\,du = u + \sum_{i=1}^{\infty} \frac{u^{2i+1}\prod_{l=1}^{l=i} l}{i!(2i+1)}$$

when $x > 1$ or $x < -1$ ie magnitude of x is greater than 1 we can use the following method

$$(1 + u)^q = u^q(1 + u^{-1})^q \quad 2.17$$

$$(1 - u)^q = u^q(u^{-1} - 1)^q \quad 2.17a$$

In the face of it we may ask why would we want to reformulate for these expressions? Why not just evaluate them directly? The point is this is an equivalent way to expand and decompose the expression and it shows the mathematical juxtaposition of the two leading to suggestions to answers to questions like $\sin^{-1}\alpha$ or $\cos^{-1}\alpha$ where $\alpha > 1$.

$(1 + u^{-1})^q$ is now able to be expanded as before.

Giving a general formula

18

$$(1+x^2)^q = x^{2q}\left(1 + \sum_{i=1}^{\infty} \frac{x^{-2i}\prod_{l=1}^{l=i} q-l+1}{i!}\right) \quad 2.18$$

$$\int (1+x^2)^q\, dx = \frac{x^{2q+1}}{2q+1} + \sum_{i=1}^{\infty} \frac{x^{2q-2i+1}\prod_{l=1}^{l=i} q-l+1}{i!\,(2q-2i+1)} \quad 2.19$$

and

$$(1-x^2)^q = -x^{2q}\left(1 + \sum_{i=1}^{\infty} \frac{(-x^{-2})^i\prod_{l=1}^{l=i} q-l+1}{i!}\right) \quad 2.18a$$

$$\int (1-x^2)^q\, dx = \frac{-x^{2q+1}}{2q+1} + \sum_{i=1}^{\infty} \frac{(-x^{-2})^i . x^{2q+1}\prod_{l=1}^{l=i} q-l+1}{i!\,(2i-2q-1)} \quad 2.19a$$

The implementation of these formulae can be made using our extended precision package. Below I set out the main program of an example of evaluating $\sin^{-1}(0.5)$ which is the practical solution of 2.09 and 2.11.

1pX2q.cpp

```
#include <math.h>
#include <cstdio>
#include <string.h>
#include <iostream>
#include <fstream>
#include <time.h>
#include <cstring>


#define ncells 2000              //remember you need extra cells for accuracy in division using Divlx

/*
*       1pX2q.cpp
* Developed by Olatunde Adeyemo ©2022
*
```

```
 * let us assume we are working to n dp through int Ndp
 */
//#define USEC_TO_SEC 1.0e-6

using namespace std;



struct NxlongR
{
          int mant[ncells]; //each cell of array mant carries 6 digits of our extended real number
          int exp;   //exp is the exponent of our first non zero digit of mant[1]
          int sgn;   //overall sign -1 or 1 or if  NxlongR==0 sgn =0
};

//ifstream InFile;   // used with  rdXlr()

#include "./fnct.ext"
/*
 * (1+U)^q = 1+sigmai((u^i . productj(q-j+1)/(i!))
 *
 * the normal implementation is  u = -+x^2
 * here we assume  0<x<1 giving
 *
 *  int((1+x^2)^q )dx= x + sigmai((x^2i+1 . productj(q-j+1)/(i! . (2i+1))
 *  int((1-x^2)^q )dx= x + sigmai((-x^2)i+1 . productj(q-j+1)/(i! . (2i+1))
 *
 * or with x >1,
 * (1+U)^q = (U(1+U^-1)) ^q = x^2q *(1+-x^-2)^q
 * int((1+x^2)^q )dx= x^(2q+1)/(2q+1)+sigmai(x^(2(q-i)+1) . productj(2q-j+1)/(i! . 2(q-i)+1)
 *
 * int((1-x^2)^q )dx= x^(1-2q)/(2q-1) +sigmai((-1^i)x^(2qi+1) . productj(j)/(i! . (2qi+1)))
 *
 */

int main(int argc, char* argv[])
{
          //FILE* OutFile;
          NxlongR one,X,x2,q,B,fct,prd,k2,termx,tIx,smx,smIx,bufi,Kx2;
          char istring[8000], pmstr[10];

          int Ndp=1000, xlt1=0,pm,ep,expo;
          long a,b;
          double Xdbl;

          double tsec;

          time_t t1,t2;
      char filename[25];

    strcpy(filename, ".//1pX2q.txt");
```

```c
FILE* OutFile = fopen(filename, (char *) &"w");
if (OutFile == NULL) {
    printf("Cannot create output files\n");
    printf("Usage: Execute native/seq_demo2 from ...");
    exit(0);
}

    time(&t1);


    itoxR(1,&one);

/*
 * got to input x, q
 * q read as a/b , a,b integers
 */
printf("\nenter value x  in the form of 2 entries,\
    \nan upto 15 digit +ve mantissa\n\
    an integer exponent\n\
    enter mantissa eg 2.592\n");

scanf(" %la",&Xdbl);
printf("\nenter exponent eg -23 or 7\n");
    scanf("%i",&expo);
    //printf("\nXdbl in %lf",Xdbl);
    X.sgn = 1;
    if (Xdbl <0)
    {
            X.sgn =-1;
            Xdbl*= -1;
    }

    while(Xdbl<1.0)
    {
            Xdbl*=10;
            ep--;
    }

    while(Xdbl>=10.0)
    {
            Xdbl/=10;
            ep++;
    }
    //printf("\nXdbl %lf",Xdbl);
    expo+=ep;
    Xdbl*=1e5;
    //printf("\nXdbl *1e5 %lf",Xdbl);
    X.mant[1]=Xdbl;
    Xdbl-=X.mant[1];
```

21

```
Xdbl*=1e6;
X.mant[2]=Xdbl;
Xdbl-=X.mant[2];
Xdbl*=1e6;
X.mant[3]=Xdbl;
X.exp=expo;

if (expo >= 0)
        xlt1=1;

//printf("\nX[1] %d\n",X.mant[1]);
/*crstrXlr(X,60,istring);
printf("\n X in\n%s",istring);*/

printf("\nenter rational value q  in the form of 2 integer entries a and b\n \
q = a/b  b may be 1,\nenter a\n");
 scanf("%ld",&a);
 printf("enter b\n");
 scanf("%ld",&b);
 itoxR(a,&q);
 itoxR(b,&B);
 Divlxr( B,q,&q,Ndp);        // ** "real" q used in (1-u)^q


printf("\nenter plus/minus, 1 for (1+x^2), -1 for (1-x^2)\n");
scanf("%d",&pm);
if(pm==1)
        sprintf(pmstr,"1+x^2");
if(pm==-1)
        sprintf(pmstr,"1-x^2");

fprintf(OutFile,"\t\t*****+++*****\n\n\n1pX2q.cpp\n\nusing x =");
crstrXlr(X,10,istring);
fprintf(OutFile,"%s",istring);



if(a*b< 0 && xlt1 ==1)
 {
        if(pm == -1)
                X.sgn *=-1;
        Divlxr( X,one,&X,Ndp);
        a=abs(a);
        b=abs(b);
        xlt1 =2;
 }

smx =tIx=one;
fct  =one;
prd = q;
smIx = X;
```

```
Kx2= one;
prodxlr(X,X,&x2,Ndp);
fprintf(OutFile,"\n\n +-x^2=");
crstrXlr(x2,10,istring);
fprintf(OutFile,"%s",istring);
fprintf(OutFile,"\n\nusing q =a/b \t a= %ld b= %ld",a,b);
fprintf(OutFile,"\n\n expansion of (%s) ^(%ld/%ld) =",pmstr,a,b );
x2.sgn =pm;
k2 =x2;int i=1;
if(xlt1 == 1)
{
        Kx2 = x2;
        while(i<=a)
        {
                prodxlr(x2,Kx2,&Kx2,Ndp);
                i++;
        }
        NRoot(Kx2,b,&Kx2,Ndp);                          //              X^2q

        Divlxr( x2,one,&x2,Ndp);
}

 i=1;
// if(xlt1 ==0 or xlt1 == 2)
while (smIx.exp -tIx.exp<Ndp)
//while(i < 10)
{
        prodxlr(k2,prd,&termx,Ndp);
        Divlxr( fct,termx,&termx,Ndp);
        addXlr(smx,termx,&smx,Ndp);

        prodxlr(X,termx,&tIx,Ndp);
        itoxR(2*i+1,&B);
        Divlxr( B,tIx,&tIx,Ndp);
        addXlr(smIx,tIx,&smIx,Ndp);
        if(i/100*100 ==i)
        {
                crstrXlr(termx,60,istring);
                printf("\nterm\n%s",istring);

                crstrXlr(tIx,60,istring);
                printf("\ntIx Int\n%s",istring);
                printf("\n\t%d",i);
        }
        i++;
        itoxR(i,&bufi);
        prodxlr(bufi,fct,&fct,Ndp);
        prodxlr(x2,k2,&k2,Ndp);

        itoxR(1-i,&bufi);
        addXlr(q,bufi,&bufi,Ndp);
```

```
                    prodxlr(bufi,prd,&prd,Ndp);
                    /*if(i/10*10 ==i)
                    {
                                crstrXlr(k2,60,istring);
                                printf("\n i  %d\n k2\n%s",i,istring);

                                crstrXlr(prd,Ndp,istring);
                                printf("\nprd Int\n%s",istring);

                    }*/
            }

            time(&t2);
            tsec=difftime(t2, t1);

            crstrXlr(smx,Ndp,istring);
            fprintf(OutFile,"\n%s",istring);

            crstrXlr(smIx,Ndp,istring);
            fprintf(OutFile,"\n\nand\n\nintegral val \n%s",istring);

            itoxR(6,&bufi);
            prodxlr(bufi,smIx,&bufi,Ndp);
            crstrXlr(bufi,Ndp,istring);
            fprintf(OutFile,"\nPI estimate\n%s",istring);

            fprintf(OutFile,"\r\n\n total compute time  %lf sec\n",tsec);

}
```

This program is a generic program 1px2q.cpp (which in terms calculates $(1-x^{2)})^{\wedge}q$ or $(1+x^{2.})^{\wedge}q$ and it's integral from 0 to x) in this case, selection for x = 0.5, q = -1/2, and the choice of $1-x^{2.}$. Inputs are shown below.

$ ./1pX2q

enter value x in the form of 2 entries,

an upto 15 digit +ve mantissa

        an integer exponent

        enter mantissa eg 2.592

0.5

enter exponent eg -23 or 7

0

enter rational value q in the form of 2 integer entries a and b

q = a/b b may be 1,

enter a

-1

enter b

2

enter plus/minus, 1 for (1+x^2), -1 for (1-x^2)

-1

Running this program with Ndp = 1000 uses processing time of 163 seconds solving pi to 983 places and with Ndp = 3000 , 7745 sec =2 hrs 9min solving pi to 2952 places. When comparing this method with using the Machin formula (Pi in the Sky and Extended Precision Calculation, Alien Pi in The Sky 2016) GregPi.cpp calculating pi to a nominal 3000 places with the same machine takes 2485s (41.4 minutes) of processing time.

This shows the Machin formula is clearly a better method for calculating pi. The speed of the Machin formula can be attributed to the simplicity of each term of the Gregory series and that we only calculate the series twice in the calculation and the largest value of x for which we calculate for is 0.2.

Gregory series $\quad \theta = x + \dfrac{\sum\limits_{i=1}^{\infty} \left(-x^2\right)^i . x}{2i+1}$

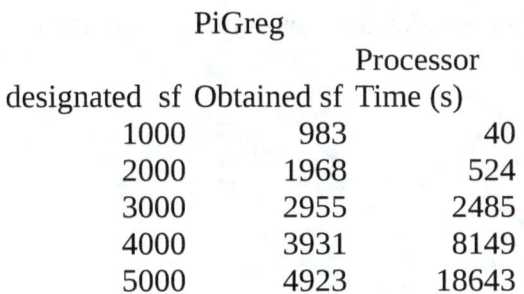

PiGreg		
designated sf	Obtained sf	Processor Time (s)
1000	983	40
2000	1968	524
3000	2955	2485
4000	3931	8149
5000	4923	18643

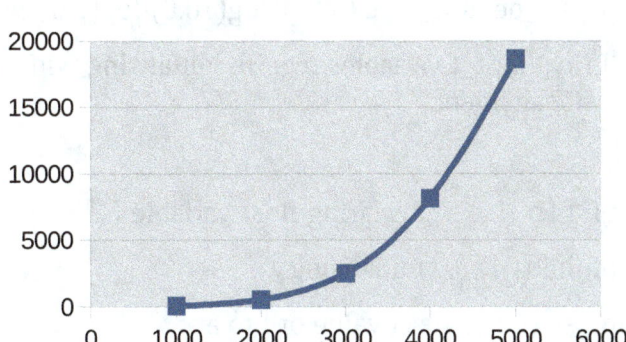

Machin formula processor time

Above the table and chart show that time of calculation increases exponentially with increase in precision. This is the result of each calculation requiring the full precision for each variable.

The exercise is still useful in demonstrating we have a method for obtaining high precision answers for integrals of series and formulae that relate to series and surds. The same method to integrate term by term may be used in calculating values for integrals of the types $x^q \sin \theta$, $x^q \cos \theta$, $x^q \sinh \theta$, $x^q \cosh \theta$, $x^q \exp(f(x))$, where q is a rational and $f(x)$ is some function of x of the form $b + a(x)^{\wedge}c$, a,b,c are constants.

GMP and arbitrary precision calculations.

GNU MP is an Opensource project (https://opensource.com/) library written in C for arbitrary precision arithmetic on integers, rational numbers, and floating-point numbers. It is freely available on the internet, details about it, explanation, download and installation are available at https://gmplib.org/. We shall use the GMP library to greatly speed up our calculations. GMP works with the architecture at low level in an optimised fashion it contains a number of extremely efficient functions for maths manipulations.

To use it the user must understand the different types of variables of the library, the main ones that we may utilise are the integer type mpz_t , the rational mpq_t and the float mpf_t. We shall mostly be concerned with the float type. We can use it to replace the NxlongR type we used in previous extended precision examples. I at first created substitute functions for the suite in fnct.ext, this was made and can be found in fns.cpp. In this way it makes adaptation of programs already created simpler and they then require the minimum of extra work. The suite of functions in fns.cpp include addXlr, prodxlr, Divlxr, crstrXlr, NRoot, itoxR, invtan, and CosSin. These functions use the GMP library to perform the same tasks as their predecessors.

In converting the main program to run with the GMP library there are a number of techniques and methods required. Initially, the precision has to be set with a set precision statement, the precision is in binary so it is convenient to convert from decimal with :-

mpf_set_default_prec (Ndp*log(10.0)/log(2.0));

All float mpf_t variables require initialising with an 'init' statement. Variable assignment is made by a 'set' statement.

Eg :-

mpf_t f;   // declaring float variable f

mpf_init(f);   // initialising

mpf_set(f,a);   // set value of f to a

All the functions of fns.cpp are to run with the default precision so unlike those of fnct.ext the extra parameter int Ndp, is removed from the input parameter list of equivalent functions. Another difference in implementing these functions is unlike in fnct.ext, those in fns.cpp often change the values of input parameters while in operation. This often requires the substitution of a "dummy variable" as the input parameter to prevent corruption. Below I show an example.

under extended precision	under GMP
	mpf_set(dup,f);
prodxlr(r5,f,&f,Ndp);	prodxlr(r5,dup,&f);

A run of the program PiGregG2.cpp, the GMP equivalent of the extended precision program GregPi.cpp calculates pi to 10,000 places in approximately 1 sec.

Below show the result of a series of runs with various precision settings.

sf (*1000)	Processor Time (s)
10	1
50	33
60	51
80	97
100	172

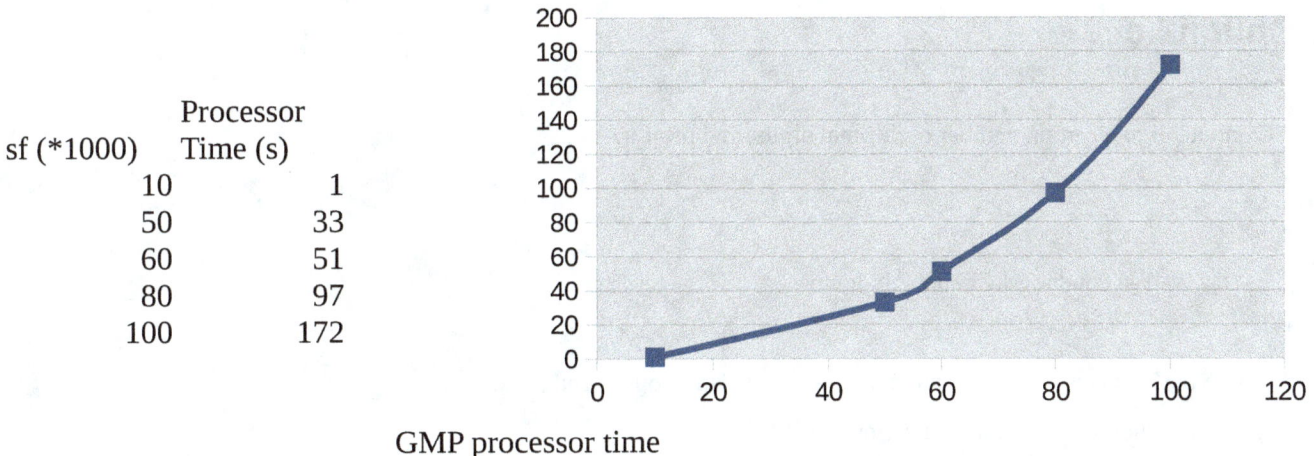

GMP processor time

Comparing with previous results from our extended precision system, that system could be projected to calculate 10,000 places taking more than 71100 seconds, really demonstrating how much faster the GMP system is.

So far the fastest formula in our estimation to calculate pi is the Machin formula. There are a number of faster formulae that have been created over the years, but notably the Chudnovsky brothers of the USA produced an unrivalled famous formula in 1987. This formula produces 14 decimal places with each term in strict succession and enables the user to easily predict its production rate.

The Chudnovsky formula.

$$\pi = \frac{426880 * \sqrt{10005}}{\sum\limits_{0}^{\infty} \dfrac{(6*k)! * (13591409 + 545140134*k)}{(3*k)! * (k!)^3 * (-640320)^{(3*k)}}}$$

A number of programs have been made using this formula. One is "chudn2.c" listed below, this runs using the GMP library and can be used to produce pi to a phenomenal number of sig fig. In an exploratory run I used it to produce a result of 100,000 sig fig using 30.389sec processing time.

For ultimate test of processing speed you can download, compile and build gmp-chudnovsky.c written by Hanhong Xue, the program is found on the GMP site, it can be used as a benchmark test. It uses the modular method of the Chudnovsky formula to divide the task into several streams that can optimise the computer architecture and run multiple threads and processors concurrently. I would not advise to print the result out unless diverted to a file however, running it for 10,000,000 sf it used a total of 5.992 sec. processing time utilising the same Intel i5-4590 CPU @ 3.30GHz × 4 architecture as in previous comparisons.

These GMP programs can be found in the GMP section of my GIT repository
https://github.com/Tunde-Adeyemo/arbitrary- p

Chudn2.c

```
/*

* Compute pi to a certain number of decimal digits, and print it.

*

*

gcc -O2 -Wall -o chudnovsky chudn2.c -lgmp

*

* WARNING: This is a demonstration program only, is not optimized for

* speed, and should not be used for serious work!

*

* The Chudnovsky Algorithm:

*                              _____

*              426880 * /10005

*         pi = --------------------------------------------

*                      _inf_

*              \    (6*k)! * (13591409 + 545140134 * k)

*               \      -----------------------------------

*               /          (3*k)! * (k!)^3 * (-640320)^(3*k)

*             /___

*                 k=0

*

* http://en.wikipedia.org/wiki/Pi#Rapidly_convergent_series

*
```

```c
#include<stdio.h>
#include<stdlib.h>
#include<gmp.h>
#include<string.h>
#include <time.h>
// how many to display if the user doesn't specify:
#define DEFAULT_DIGITS 60
// how many decimal digits the algorithm generates per iteration:
#define DIGITS_PER_ITERATION 14.1816474627254776555
/**
 * Compute pi to the specified number of decimal digits using the
 * Chudnovsky Algorithm.
```

```
 *

 * http://en.wikipedia.org/wiki/Pi#Rapidly_convergent_series

 *

 * NOTE: this function returns a malloc()'d string!

 *

 * @param digits number of decimal digits to compute

 *

 * @return a malloc'd string result (with no decimal marker)

 */

char *chudnovsky(unsigned long digits)

{

          mpf_t result, con, A, B, F, sum;

          mpz_t a, b, c, d, e;

          char *output;

          mp_exp_t exp;

          double bits_per_digit;

          unsigned long int k, threek;

          unsigned long iterations = (digits/DIGITS_PER_ITERATION)+1;

          unsigned long precision_bits;

          // roughly compute how many bits of precision we need for

          // this many digit:

          bits_per_digit = 3.32192809488736234789; // log2(10)

          precision_bits = (digits * bits_per_digit) + 1;

          mpf_set_default_prec(precision_bits);

          // allocate GMP variables

          mpf_inits(result, con, A, B, F, sum, NULL);

          mpz_inits(a, b, c, d, e, NULL);

          mpf_set_ui(sum, 0); // sum already zero at this point, so just FYI

          // first the constant sqrt part

          mpf_sqrt_ui(con, 10005);

          mpf_mul_ui(con, con, 426880);

          // now the fun bit

          for (k = 0; k < iterations; k++)

          {

                    threek = 3*k;
```

```c
        mpz_fac_ui(a, 6*k);
        // (6k)!
        mpz_set_ui(b, 545140134); // 13591409 + 545140134k
        mpz_mul_ui(b, b, k);
        mpz_add_ui(b, b, 13591409);
        mpz_fac_ui(c, threek);
        // (3k)!
        mpz_fac_ui(d, k); // (k!)^3
        mpz_pow_ui(d, d, 3);
        mpz_ui_pow_ui(e, 640320, threek); // -640320^(3k)
        if ((threek&1) == 1) { mpz_neg(e, e); }
        // numerator (in A)
        mpz_mul(a, a, b);
        mpf_set_z(A, a);
        // denominator (in B)
        mpz_mul(c, c, d);
        mpz_mul(c, c, e);
        mpf_set_z(B, c);
        // result
        mpf_div(F, A, B);
        // add on to sum
        mpf_add(sum, sum, F);
    }
    // final calculations (solve for pi)
    mpf_ui_div(sum, 1, sum); // invert result
    mpf_mul(sum, sum, con); // multiply by constant sqrt part
    // get result base-10 in a string:
    output = mpf_get_str(NULL, &exp, 10, digits, sum); // calls malloc()
    // free GMP variablesmpf_clears(result, con, A, B, F, sum, NULL);
    mpz_clears(a, b, c, d, e, NULL);
    return output;
}
/**
 * Print a usage message and exit
 */
```

```c
void usage_exit(void)
{
        fprintf(stderr, "usage: chudnovsky [digits]\n");
        exit(1);
}
cputime ()
{
  return (int) ((double) clock () * 1000 / CLOCKS_PER_SEC);
}
/**
* MAIN
*
* See usage_exit() for usage.
*/
int main(int argc, char **argv)
{
        char *pi, *endptr;
        long digits;
        long start, end;
        switch (argc) {
                case 1:
                digits = DEFAULT_DIGITS;
                break;
                case 2:
                digits = strtol(argv[1], &endptr, 10);
                if (*endptr != '\0') { usage_exit(); }
                break;
                default:
                usage_exit();
        }
        if (digits < 1) { usage_exit(); }
        start = cputime();
        pi = chudnovsky(digits);
        end = cputime();
        // since there's no decimal point in the result, we'll print the
```

```
// first digit, then the rest of it, with the expectation that the
// decimal will appear after "3", as per usual:
printf("%.1s.%s\n", pi, pi+1);  //save the screen
printf("time to calc %ld sig fig = %6.3f\n",digits, (double)(end-start)/1000);
// chudnovsky() malloc()s the result string, so let's be proper:
free(pi);
return 0;
}
```

Improved Root Iterations

$$f(x)=x^n-A=0$$

given

x_1 such that $f(x_1)=f(x+\varepsilon), \varepsilon < x$

$$f(x_1) \approx f(x)+\varepsilon f'(x)+\varepsilon^2 \frac{f^2(x)}{2} \ldots +\varepsilon^n \frac{f^n(x)}{n!}$$

take first 2 terms and $\quad f(x)=x^n-A=0$

$$f(x_1)=f(x)+\varepsilon f'(x)$$

$$f(x_1)=0+n\varepsilon x^{n-1}, x_1=x+\varepsilon$$

$$\varepsilon = \frac{f(x_1)}{n x^{n-1}}$$

if

$$\varepsilon = \epsilon x, \epsilon = \frac{f(x_1)}{nA}$$

$$x_1=x(1+\epsilon) \qquad\qquad x=\frac{x_1}{1+\dfrac{f(x_1)}{nA}}=\frac{nA x_1}{nA+f(x_1)}=\frac{nA x_1}{nA+x_1^n-A}$$

This is the formula used in improved implementation of root iterations in the function NRoot() given below.

```
void NRoot(NxlongR A, int N,NxlongR* x,int Ndp)  //05/22   /uses
// x(n+1)=ANx(n)/(AN-y) , y = x(n)^N -A
{
        NxlongR one,y,na;
        int i, i2=1,ye= A.exp,aN;
        char istring[100];
        Ndp*=1.2;



        aN=N;
        itoxR(1,&one);
        if (N < 0)
        {
                printf("\nN is %d\n",N);
                Divlxr( A,one,&A,1.2*Ndp);
                crstrXlr(A,15,istring);
                printf("\n 1/A      %s ",istring);
                aN *= -1;
        }

        itoxR(aN,&na);
        *x=y=one;

  if((A.exp==-1 && A.mant[1]<800000) ||A.exp<-1 )
   {
                  Divlxr( na,A,&y,Ndp);
     if(A.exp<-5)
                {
                    y.exp=(A.exp-aN+1)/aN;
                }

    *x=y;
        }
        prodxlr(*x, *x,&y,Ndp);
        for(i=3;i<=aN;++i)
        {
                prodxlr(*x, y,&y,Ndp);
        }
        prodxlr(na, A,&na,Ndp);
        A.sgn*=-1;     // -A


        addXlr(y,A,&y,Ndp);

        addXlr(y,na,&y,Ndp);
        Divlxr( y,*x,&y,Ndp);
```

```
        prodxlr(na, y,x,Ndp);

        while(A.exp-ye< 0.9*Ndp )
        {
                prodxlr(*x, *x,&y,Ndp);
                for(i=3;i<=aN;++i)
                {
                        prodxlr(*x, y,&y,Ndp);
                }
                addXlr(y,A,&y,Ndp);

                if (y.sgn == 0)
                        break;
                ye = y.exp;

                addXlr(y,na,&y,Ndp);

                Divlxr( y,*x,&y,Ndp);

                prodxlr(na, y,x,Ndp);
                i2++;
        }
        if(N<0)
                printf("\nN is %d\n",N);
        return;
}
```

Regular Polyhedra

Let us examine regular convex polyhedra. These are polyhedra based on a sphere that have facets (faces) that are regular (congruent) polygons and corners (vertices) that form a regular shape that looks the same from any direction. This means all the faces are equivalent and all the vertices are equivalent.

Structurally, spars equivalent to edges of these regular polyhedra make the best solutions for evenly distributing compressive or expansive loads over a spherical surface and as such they have importance in structural design and the regularity and simplicity offers an additional aesthetic pleasure.

There are 5 Platonic solids that meet these conditions, they have been known since ancient times, they were taught in Plato's schools of philosophy before 360 BC. The Platonic Solids are the tetrahedron (4 faces, 4 vertices), the cube (6 faces, 8 vertices), the octahedron(8 faces, 6 vertices), the dodecahedron (12 faces, 20 vertices)and the icosahedron (20 faces, 12 vertices). They obey Euler rule :

faces+vertices – edges= 2   or written more commonly   V- E + F =2.

They can be described by their Schläfli symbol represented {m, n} where m is the number of edges per face and n is the number of faces that meet at a vertex.

	Schläfli symbol
tetragon	{3, 3}
cube	{4, 3}
octagon	{3, 4}
dodecahedron	{5, 3}
icosahedron	{3, 5}

Let us see how we can generate them as 3d figures.

The point is for a 3d vertex the minimum number of edges that can meet at the vertex is 3. On a plane three 120° equiangular angles form a plane and this would be the external angle of a 6 sided regular polygon. In order that our polyhedra to remain convex the external angles have got to be less than 120° and the maximum size regular polygon we can use is therefore limited to 5 sided.

A similar argument is held to show that the maximum number of edges meeting at a vertex is 5 from a polygon with 3 sides ie a an equilateral triangle. The minimum angle of a regular polygon is 60° and similarly 6 of them would form a plane and the figure would not be convex.

We are left with polygons with only 3, 4 or 5 sides to create our regular convex polyhedra. With a polygon with 3 sides we can create a polyhedra with 3, 4 or 5 edges per vertex, with a polygon with 4 or 5 sides we may only produce a regular convex polyhedron with 3 edges per vertex.

Having decided the regular polygon and number of edges per vertex, we need to be to find a transformation that relates adjacent faces, ie. what angle should a face to be rotated about its edge to map it onto it's neighbour.

To consider a tetragon we start with an equilateral triangle. The equilateral triangle in the z = 0 plane with points

x	y	z
0	0	0
1	0	0
0.5	0.866025	0
0	0	0

Each side has side length 1.0 .
To create a tetragon, we need to define a 4th point that meets the criteria of being equidistant from each point the triangle. We may consider rotating the triangle about an edge.

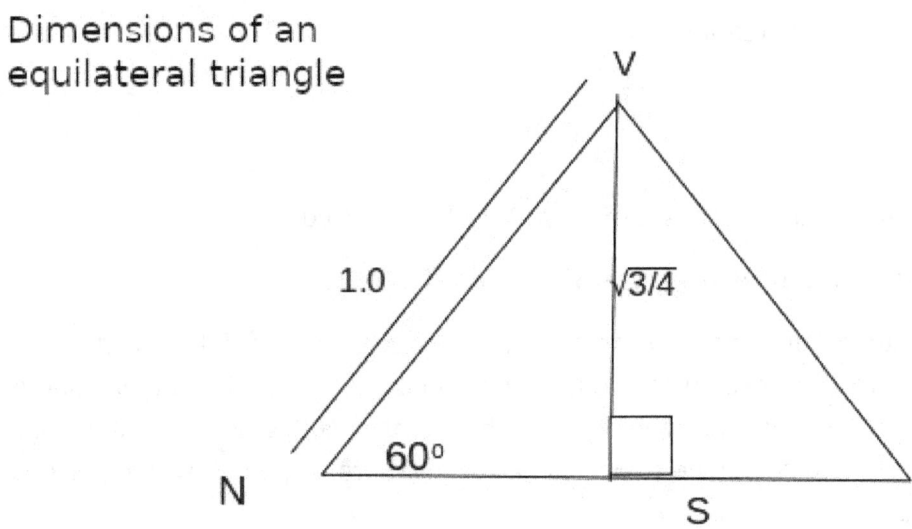

Dimensions of an equilateral triangle

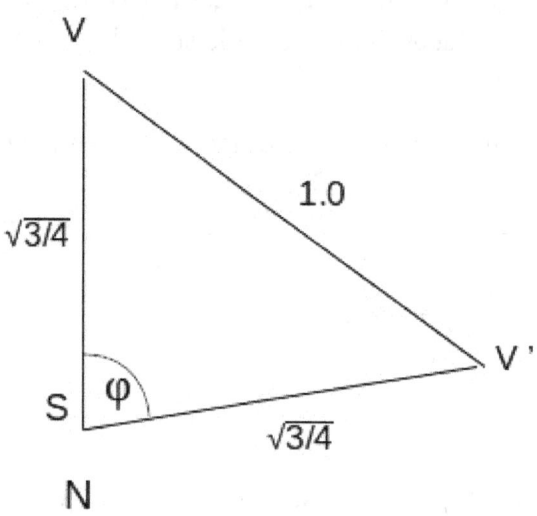

Angle of rotation for 4th point of Tetrahedron

To find φ consider the rotation about axis side S from V to V' observed from N.

It is easily shown $φ/2 = \sin{-1}(0.5/ \sqrt{(3/4)}) = 35.26°$, $φ = 70.56°$.

In "Matrix Magic", Olatunde Adeyemo, 2010 in section 3, "Transformations" I created a Microsoft Windows application UniaxialRot.exe . It would transform a lattice of points into their rotation through a prescribed angle about an axis running through a point. The engine of this application is useful to us here and I used it to create the C++ program "uniaxialrot.cpp" below that performs the same task in the same fashion.

When compiled and built in it's own directory, it requires an input file " inrota.txt" listing the details of rotation and points to be transformed. It should lie within the directory of the exec file.

uniaxialrot.cpp

```cpp
#include <math.h>
#include <stdio.h>
#include <string.h>
#include <iostream>
#include<fstream>
#include <time.h>
#include <cstring>

/*Developed by Olatunde Adeyemo mar 2021
uniaxrot.cpp is the cpp version of the c# visual studio program UniaxialRot

*/

using namespace std;


double dot(double va[],double vb[])
{
        double z=0.0;
        for(int i=1;i<4;i++)
                z+=va[i]*vb[i];
        return z;
}

void cross(double va[],double vb[],double *perp)
{
        perp[1]=va[2]*vb[3]-va[3]*vb[2];
        perp[2]=va[3]*vb[1]-va[1]*vb[3];
        perp[3]=va[1]*vb[2]-va[2]*vb[1];

}

double fnorm(double va[])
{
        double z=0.0;
        for(int i=1;i<4;i++)
                z+=va[i]*va[i];

        z=sqrt(z);

        return z;
}

void rotret(double xyz[],double p[], double Vb[],double theta, char* Os)
{
        char var[30];
        double xP,yP,zP,parlc,ts, trad=2*theta*asin(1.0)/180,dp;
        double Va[4],Vc[4],Vo[4];
        int i;
        for(i=1;i<=3;i++)
        Va[i]=xyz[i]-p[i];              //Va a -> P(x,y,z)
        Va[0]=fnorm(Va);
        //printf("\n norm %lf",Va[0]);
```

```c
        Vb[0]=fnorm(Vb);
        for(i=1;i<=3;i++)
        Vo[i]=Vb[i]/Vb[0];


        parlc=dot(Va,Vo);
        xP=parlc*Vo[1];//xyzP comp a->P parll to Vxyz
        yP=parlc*Vo[2];
        zP=parlc*Vo[3];

        Va[1]-=xP;          //Va comp a->P perp toVxyz
        Va[2]-=yP;
        Va[3]-=zP;
        /*printf("\nVa prp \t");
        for(i=1;i<=3;i++)
        printf("%lf \t",Va[i]);


        cross(Va,Vo,Vc);
        printf("\nVc \t");
        for(i=1;i<=3;i++)
        printf("%lf \t",Vc[i]);

        printf("\n\n");*/

        cross(Va,Vo,Vc);

        dp=cos(trad);
        ts=sin(trad);
        //printf("dp = %lf ts =%lf\n",dp,ts);


        Vo[1]=p[1]+xP+dp*Va[1]-ts*Vc[1];
        Vo[2]=p[2]+yP+dp*Va[2]-ts*Vc[2];
        Vo[3]=p[3]+zP+dp*Va[3]-ts*Vc[3];
        //printf("zP = %lf \tVa[3]=%lf \tVc[3]=%lf\n",zP,Va[3],Vc[3]);

        strcpy(Os,"");
        for(i=1;i<=3;i++)
        {
                sprintf(var,"%lf \t",xyz[i]);
                strcat(Os, var);
        }
        for(i=1;i<=3;i++)
        {
                sprintf(var,"%lf \t",Vo[i]);
                strcat(Os,var);
        }
        strcat(Os,"\n");

}

int main(int argc, char* argv[])
{
        int i;
        double p[4],v[4],xyz[4],theta;
```

```
ifstream InFile;

char filename[25], istring[100],Os[200];

strcpy(filename, ".//inrota.txt");
InFile.open(filename, ios::in);

for(i=1;i<=6;i++)
InFile.getline(istring,100,'\n');

sscanf(istring,"point     px      py      pz %lf   %lf %lf",&p[1],&p[2],&p[3]);
InFile.getline(istring,100,'\n');
sscanf(istring,"vector    vx      vy      vz %lf   %lf %lf",&v[1],&v[2],&v[3]);
InFile.getline(istring,100,'\n');
sscanf(istring,"theta \t%lf",&theta);


strcpy(filename, ".//ofrota.txt");

ofstream OutFile(filename, ios::out);

sprintf(Os,"\n\t\tresult of Rotation of %lf deg \n\t\
about vector %lf %lf %lf\n\t passing through point %lf %lf %lf\n\n\n",theta,v[1],v[2],v[3],\
p[1],p[2],p[3]);
OutFile<< Os;

for(i=1;i<=3;i++)
InFile.getline(istring,100,'\n');

while(InFile.eof() == 0)
{
        InFile.getline(istring,100,'\n');
        i=sscanf(istring,"%lf %lf %lf",&xyz[1],&xyz[2],&xyz[3]);
        //cout<< istring<<"\n";
        if (i>0)
        {
                rotret( xyz,p , v,theta, Os);
                OutFile<<Os;
        }
}

}
```

Using this application on our equilateral triangle and it rotating through φ =70.56° about an edge should meet our conditions. We may choose to do this about the edge that lies between points (0,0,0) and (1,0,0). To accomplish this the input file "inrota.txt" is edited to the form below.

inrota.txt

template for input

point	px	py	pz	1.0	0.0	0.0
vector	vx	vy	vz	1.0	0.0	0
theta	70.5288		deg			

data

x	y	z
0	0	0
1	0	0
0.5	0.866025	0
0	0	0

Running uniaxialrot with this as the input file will produce output file "ofrota.txt" with content below.

ofrota.txt

result of Rotation of 70.528779 deg
about vector 1.000000 0.000000 0.000000
passing through point 1.000000 0.000000 0.000000

0	0	0	0	0	0
1	0	0	1	0	0
0.5	0.866025	0	0.5	0.288675	0.816496
0	0	0	0	0	0

Since the points (0,0,0) and (1,0,0) are both on the axis of rotation they remain invariant. Only the point (0.5, 0.866025, 0) is transformed to (0.500000, 0.288675, 0.816496). This the 4th point of our tetrahedron, to compose the full figure we must join all the points and it may be described :

0	0	0
1	0	0
0.5	0.866025	0
0	0	0
0.5	0.288675	0.816496
1	0	0
0.5	0.288675	0.816496
0.5	0.866025	0

The average of the 4 points is shown to be

0.5	0.288675	0.20412425

To centre the figure we subtract this from each point to give the result.

-0.5	-0.288675	-0.20412425
0.5	-0.288675	-0.20412425
0	0.57735	-0.20412425
-0.5	-0.288675	-0.20412425
0	0	0.61237275
0.5	-0.288675	-0.20412425
0	0	0.61237275
0	0.57735	-0.20412425

To visualise this we can copy these coordinates to a spreadsheet and create charts of XY plane and XZ plane, however because of the symmetry in the X axis we could give the figure a slight rotation about the Y axis to give what looks like the 3 dimensional figure.

result of Rotation of 45.000000 deg
about vector 0.000000 1.000000 0.000000
passing through point 0.000000 0.000000 0.000000

-0.5	-0.288675	-0.204124	-0.497891	-0.288675	0.209216
0.5	-0.288675	-0.204124	0.209216	-0.288675	-0.497891
0	0.57735	-0.204124	-0.144338	0.57735	-0.144338
-0.5	-0.288675	-0.204124	-0.497891	-0.288675	0.209216
0	0	0.612373	0.433013	0	0.433013
0.5	-0.288675	-0.204124	0.209216	-0.288675	-0.497891
0	0	0.612373	0.433013	0	0.433013
0	0.57735	-0.204124	-0.144338	0.57735	-0.144338
0	0.57735	-0.204124	-0.176777	0.57735	-0.102062

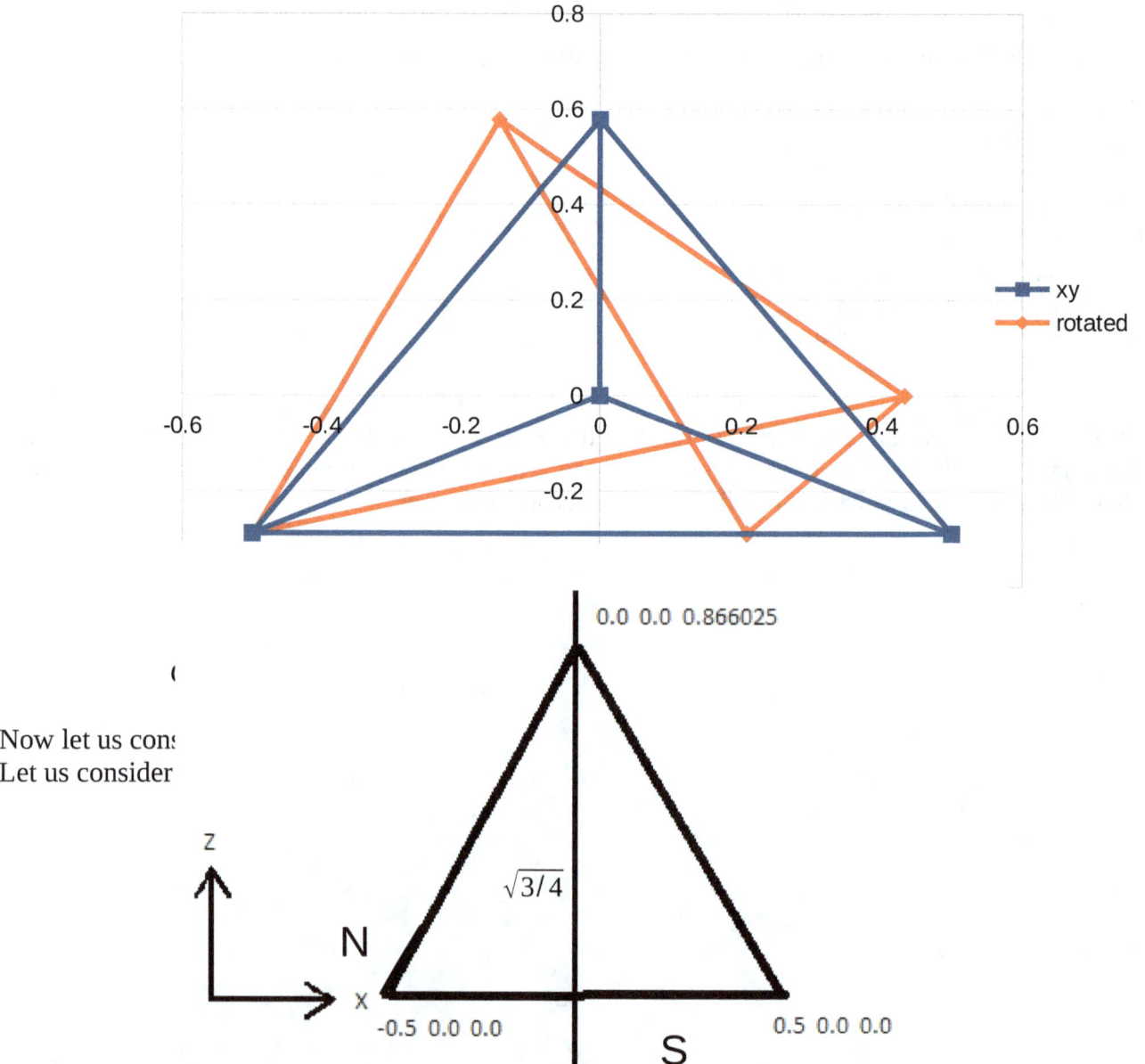

Now let us con:
Let us consider

Face of octahedron

We seek a rotation about side S to create a new face of our figure.

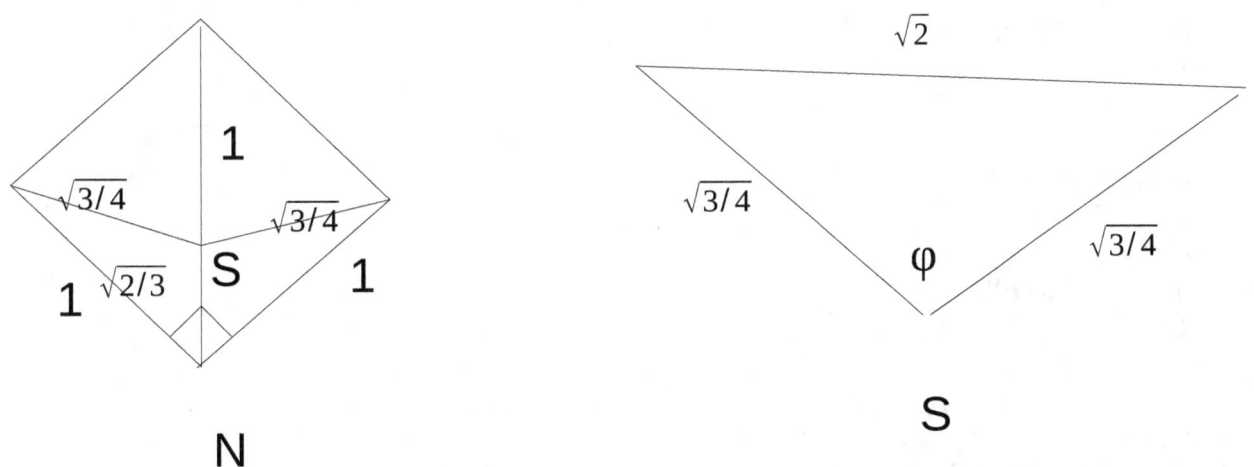

Angle between the faces of an octahedron

It can be shown φ = 2 sin⁻¹($\sqrt{(2/3)}$) = 2*54.736°. To create a kind of symmetry we can rotate our initial face by 90° - 54.736° about it's bottom edge, then we rotate the new face by 109.471°.

The resulting two faces written as point coordinates:

```
-0.5    0       0
0       -0.5    0.707107
0.5     0       0
-0.5    0       0


-0.5    0       0
0       -0.5    -0.707106
0.5     0       0
-0.5    0       0
```

Now because we have created them symmetric in Z axis we can rotate it 90° ,180° and 270° about an axis in Z direction passing point (0   -0.5   0.707107) to create the next 6 faces.

```
-0.5    -1      0
0       -0.5    0.707107
-0.5    0       0
-0.5    -1      0
0       -0.5    -0.707106
-0.5    0       0
-0.5    -1      0
0.5     -1      0
0       -0.5    0.707107
-0.5    -1      0
0.5     -1      0
0       -0.5    -0.707106
-0.5    -1      0
0.5     -1      0
0.5     0       0
0       -0.5    0.707107
0.5     -1      0
0.5     0       0
0       -0.5    -0.707106
0.5     -1      0
0.5     0       0
```

The six vertices of the figure are

```
-0.5            0               0
0               -0.5    0.707107
0.5             0               0
0               -0.5    -0.707106
0.5             -1              0
-0.5            -1              0
```

with an average position (0, -0.5, 0), subtracting this from all the points of the complete figure centres it on the origin giving:

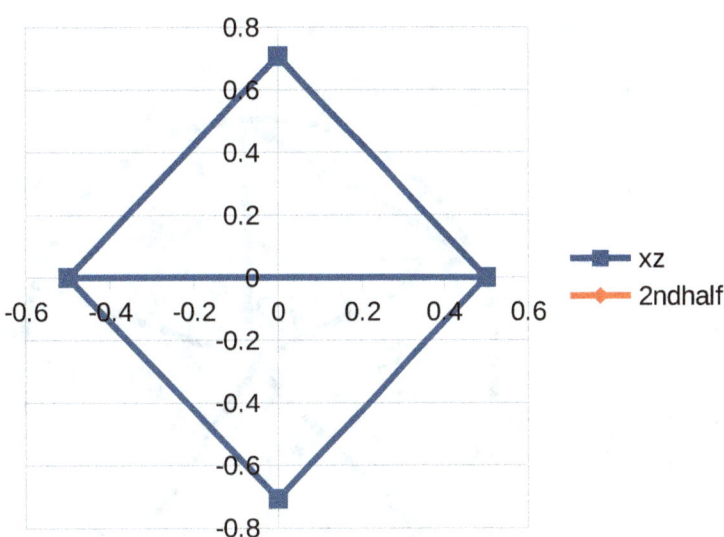

-0.5	0.5	0
0	0	0.707107
0.5	0.5	0
-0.5	0.5	0
0	0	-0.707106
0.5	0.5	0
-0.5	0.5	0
-0.5	-0.5	0
0	0	0.707107
-0.5	0.5	0
-0.5	-0.5	0
0	0	-0.707106
-0.5	0.5	0
-0.5	-0.5	0
0.5	-0.5	0
0	0	0.707107
-0.5	-0.5	0
0.5	-0.5	0
0	0	-0.707106
-0.5	-0.5	0
0.5	-0.5	0
0.5	0.5	0
0	0	0.707107
0.5	-0.5	0
0.5	0.5	0
0	0	-0.707106
0.5	-0.5	0
0.5	0.5	0

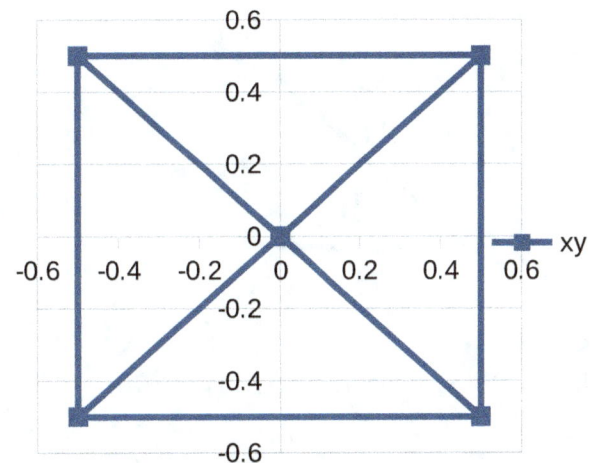

This forms our complete octahedron which can viewed if we transfer them to a spreadsheet and draw a chart.

There are a number of 3d tools we can use to visualise our complete figure including adapting it to be visualised on OpenGL or the beta graphics package Processing.com.

The final regular polyhedron that uses an equilateral triangle as a base face is the icosahedron, this form has 5 faces meeting at a vertex. Here we show the view directly above a vertex in plan.

Plan view of an icosahedron

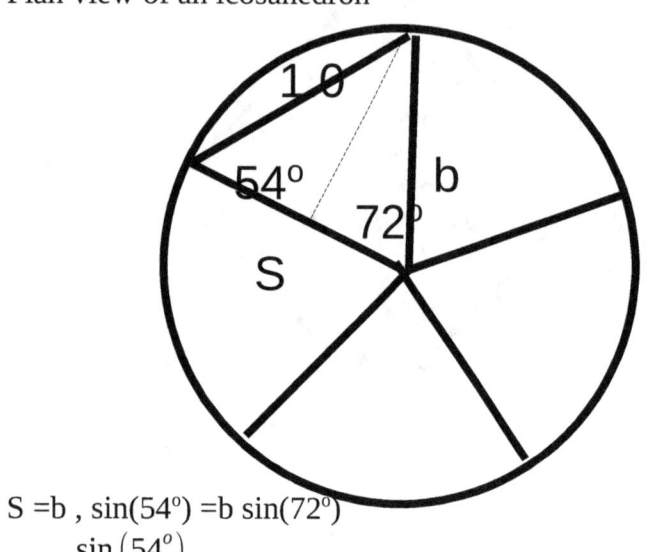

XY

$S = b$, $\sin(54°) = b\sin(72°)$

$$b = \frac{\sin(54°)}{\sin(72°)} = 0.85065$$

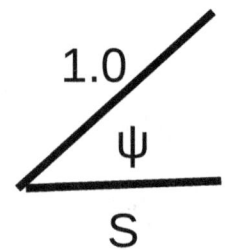

$\psi = \cos^{-1}(S) = \cos^{-1}(b) = 31.717°$

We may centre the vertex at (0.0, 0.0, sin(31.717)) = (0.0, 0.0, 0.5257). Our first equilateral may be described

0.0,	0.0,	0.5257
0.0,	0.85065,	0.0
-0.80902,	0.26287,	0

Using the geometries we find the angle relating adjacent faces.

S

$\sqrt{3/4}$ φ $\sqrt{3/4}$ 48

$$\varphi = 2\sin^{-1}\left(\frac{\sin 54^{o}}{\sqrt{3/4}}\right) = 138.1897^{o}$$

We can generate the full vertex by rotating this triangle about the Z axis by 72°, 144°, 216° and 288° . Resulting in an array describing it:

0.00000	0.00000	0.52573
0.00000	0.85065	0.00000
-0.80902	0.26287	0.00000
0.00000	0.85065	0.00000
0.80902	0.26287	0.00000
0.00000	0.00000	0.52573
0.80902	0.26287	0.00000
0.50000	-0.68819	0.00000
0.00000	0.00000	0.52573
0.50000	-0.68819	0.00000
-0.50000	-0.68819	0.00000
0.00000	0.00000	0.52573
-0.80902	0.26287	0.00000
-0.50000	-0.68819	0.00000

Our next task is to generate the next ring of faces. This can be done by taking each triangle of the vertex and doing the appropriate rotation about it its external edge, e.g. the first triangle

0.00000	0.00000	0.52573
0.00000	0.85065	0.00000
-0.80902	0.26287	0.00000
0.00000	0.85065	0.00000

will be rotated by 138.1897° about the edge

0.00000	0.85065	0.00000
-0.80902	0.26287	0.00000

We can edit inrota.txt to

template for input

point	px	py	pz	0.00000	0.85065	0.00000
vector	vx	vy	vz	0.80902	0.58779	0
theta	138.18969	deg				

data

x	y	z
0.00000	0.00000	0.52573
0.00000	0.85065	0.00000
-0.80902	0.26287	0.00000
0.00000	0.85065	0.00000

This produces the new face

-0.5	0.688188	-0.850649
0	0.85065	0
-0.809012	0.262859	0
0	0.85065	0

Rearranging the new point for convenience we can reduce the triangle to 3 points

-0.809012	0.262859	0
-0.5	0.688188	-0.850649
0	0.85065	0

We can again use the symmetry of the Z axis to produce the next 4 faces of the ring by rotating this triangle about the Z axis by 72°, 144°, 216° and 288°.

-0.809016	0.262865	0
-0.809014	-0.262866	-0.850649
-0.499992	-0.688188	0
-0.5	-0.68819	0
0	-0.850648	-0.850649
0.5	-0.688183	0
0.5	-0.68819	0
0.809015	-0.262863	-0.850649
0.809009	0.262868	0
0.809016	0.262865	0

| 0.499997 | 0.68819 | -0.850649 |
| 0 | 0.850644 | 0 |

Plotting the points out reveal we have produced a hemisphere.

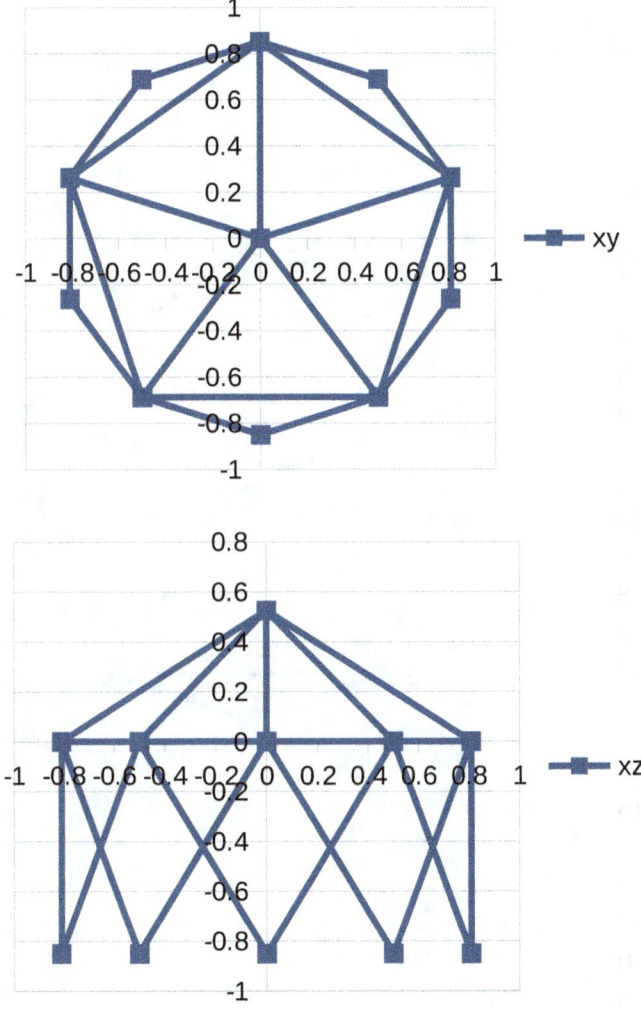

Top hemisphere of icosahedron

We find the central band is made 10 points. Taking the average of these will give the centre of the structure.

-0.809016		
	0.262865	0
-0.5	-0.68819	0
0.5	-0.688183	0
0.809016	0.262865	0
0	0.850651	0
-0.809014	-0.26	-0.85
0	-0.85	-0.85
0.809015	-0.26	-0.85
0.499997	0.69	-0.85
-0.5	0.69	-0.85

Average

0	0	-0.4253245

Now a rotation of 180° about vector (1,0,0) passing through the central point will produce the second hemisphere.

0	0	-1.37638
0	-0.850651	-0.850649
-0.809017	-0.262866	-0.850649
0	-0.850651	-0.850649
0.809017	-0.262866	-0.850649
0	0	-1.37638
0.809017	-0.262866	-0.850649
0.500001	0.68819	-0.850649
0	0	-1.37638
0.500001	0.68819	-0.850649
-0.5	0.688191	-0.850649
0	0	-1.37638
-0.809017	-0.262865	-0.850649
-0.5	0.688191	-0.850649
-0.809012	-0.262859	-0.850654
-0.5	-0.688188	0
0	-0.85065	-0.850649
-0.809016	-0.262865	-0.850649
-0.809014	0.262866	0
-0.499992	0.688188	-0.850649
-0.5	0.68819	-0.850649
2E-06	0.850648	0
0.5	0.688183	-0.850649
0.5	0.68819	-0.850649
0.809015	0.262863	0
0.809009	-0.262868	-0.850649
0.809016	-0.262865	-0.850649
0.499997	-0.68819	0
-5E-06	-0.850644	-0.850649

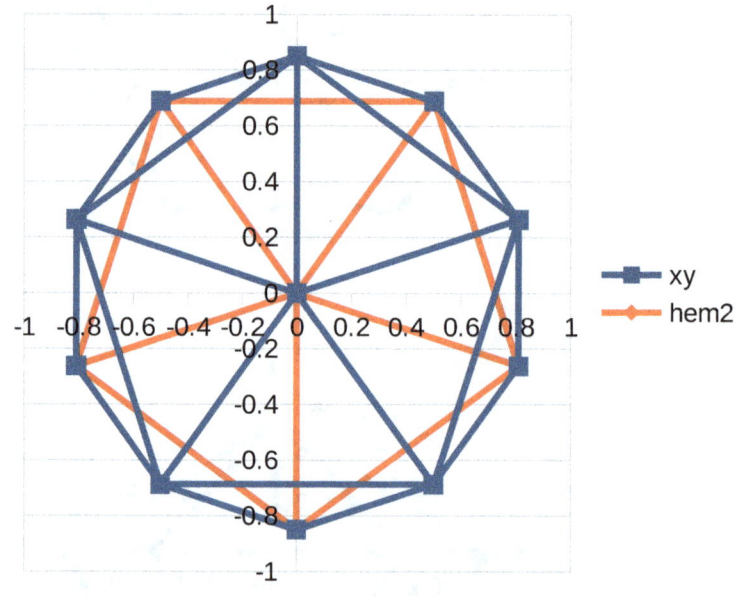

The final frame can be centred by subtracting the average position from each point.

0.0000	0.0000	0.9511
0.0000	0.8507	0.4253
-0.8090	0.2629	0.4253
0.0000	0.8507	0.4253
0.8090	0.2629	0.4253
0.0000	0.0000	0.9511
0.8090	0.2629	0.4253
0.5000	-0.6882	0.4253
0.0000	0.0000	0.9511
0.5000	-0.6882	0.4253
-0.5000	-0.6882	0.4253
0.0000	0.0000	0.9511
-0.8090	0.2629	0.4253
-0.5000	-0.6882	0.4253
-0.8090	0.2629	0.4253
-0.5000	0.6882	-0.4253
0.0000	0.8506	0.4253
-0.8090	0.2629	0.4253
-0.8090	-0.2629	-0.4253
-0.5000	-0.6882	0.4253
-0.5000	-0.6882	0.4253
0.0000	-0.8506	-0.4253
0.5000	-0.6882	0.4253
0.5000	-0.6882	0.4253
0.8090	-0.2629	-0.4253
0.8090	0.2629	0.4253
0.8090	0.2629	0.4253
0.0000	0.8506	0.4253
0.5000	0.6882	-0.4253
0.0000	0.0000	-0.9511
0.0000	-0.8507	-0.4253
-0.8090	-0.2629	-0.4253
0.0000	-0.8507	-0.4253
0.8090	-0.2629	-0.4253
0.0000	0.0000	-0.9511
0.8090	-0.2629	-0.4253
0.5000	0.6882	-0.4253
0.0000	0.0000	-0.9511
0.5000	0.6882	-0.4253
-0.5000	0.6882	-0.4253
0.0000	0.0000	-0.9511

-0.8090	-0.2629	-0.4253
-0.5000	0.6882	-0.4253
-0.8090	-0.2629	-0.4253
-0.5000	-0.6882	0.4253
0.0000	-0.8507	-0.4253
-0.8090	-0.2629	-0.4253
-0.8090	0.2629	0.4253
-0.5000	0.6882	-0.4253
-0.5000	0.6882	-0.4253
0.0000	0.8506	0.4253
0.5000	0.6882	-0.4253
0.5000	0.6882	-0.4253
0.8090	0.2629	0.4253
0.8090	-0.2629	-0.4253
0.8090	-0.2629	-0.4253
0.5000	-0.6882	0.4253
0.0000	-0.8506	-0.4253

We all know the cube is the polyhedron based on the square. It is elementary to build a cube. Below I set out the coordinates and I show what it looks like rotated 20° about (1,1,1).

0.5	0.5	-0.5
0.5	-0.5	-0.5
-0.5	-0.5	-0.5
-0.5	0.5	-0.5
-0.5	0.5	0.5
-0.5	-0.5	0.5
-0.5	-0.5	-0.5
0.5	-0.5	-0.5
0.5	-0.5	0.5
0.5	0.5	0.5
0.5	0.5	-0.5
-0.5	0.5	-0.5
-0.5	0.5	0.5
0.5	0.5	0.5
0.5	-0.5	0.5
-0.5	-0.5	0.5

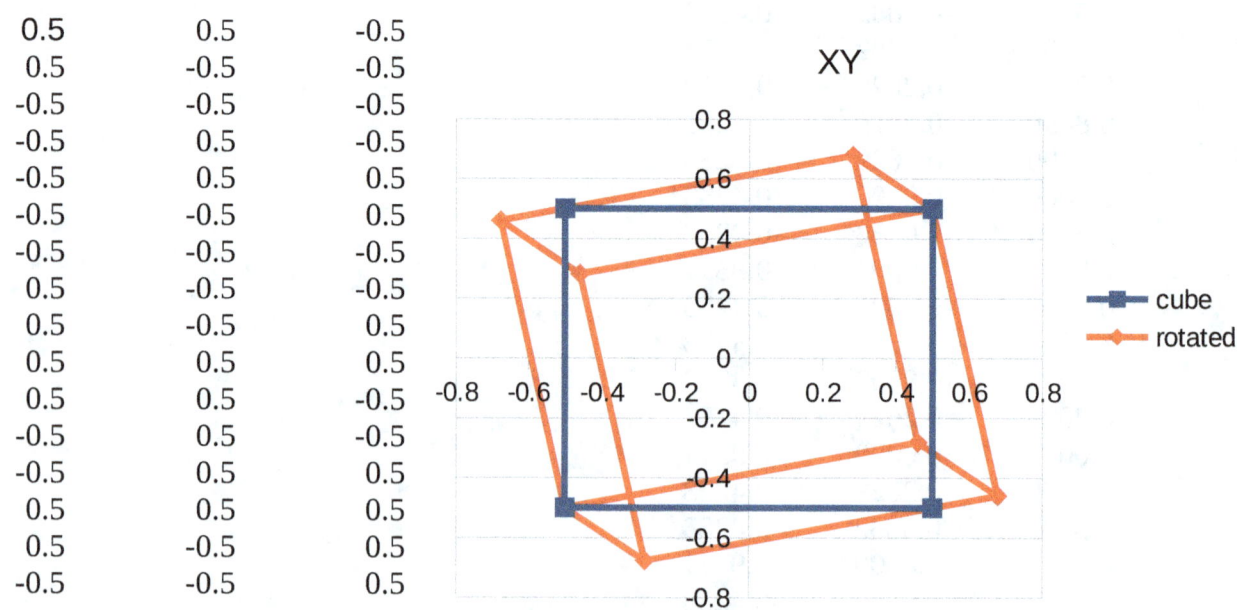

The dodecahedron is the only regular polyhedron that uses a pentagon as it's base polygon. Given each edge has a nominal 1 unit length, we set out important dimensions.

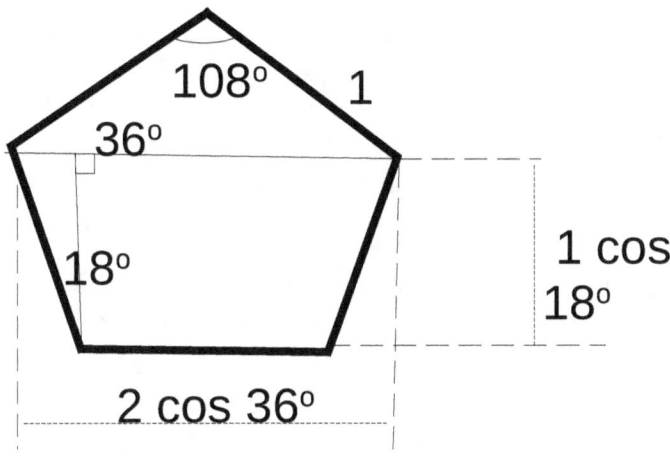

Dimensions of a regular pentagon

Again we aim to find what angle we need to tilt from one face to the next.

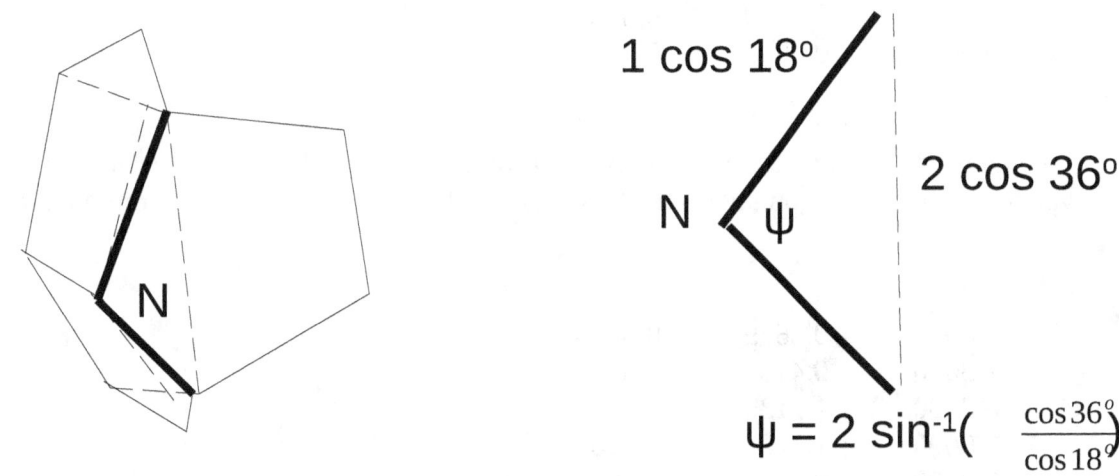

$$\psi = 2 \sin^{-1}\left(\frac{\cos 36^{\circ}}{\cos 18^{\circ}}\right)$$

Angle between faces of dodecahedron

This gives a value for $\psi = 116.5651^{\circ}$. Let us create an origin centred pentagon in the XY plane with sides of length 1.0.

0.80902	0.26287	0.00000
0.00000	0.85065	0.00000
-0.80902	0.26287	0.00000
-0.50000	-0.68819	0.00000
0.50000	-0.68819	0.00000
0.80902	0.26287	0.00000

This can be rotated through an edge by ψ. If we choose the edge with points

0.80902	0.26287	0.00000
0.00000	0.85065	0.00000

giving pentagon B:

0.8090	0.2629	0.0000
0.0000	0.8507	0.0000
0.0000	1.3764	-0.8507
0.8090	1.1135	-1.3764
1.3090	0.4253	-0.8507
0.8090	0.2629	0.0000
0.0000	0.8507	0.0000

We can use the symmetry of the Z axis to produce the next 4 faces of the ring by rotating an edited version of this pentagon about the Z axis by 72°, 144°, 216° and 288° to complete a hemisphere.

-0.8090	0.2629	0.0000
-1.3090	0.4253	-0.8507
-0.8090	1.1135	-1.3764
0.0000	1.3764	-0.8507
0.0000	0.8507	0.0000
-0.8090	0.2629	0.0000
-0.5000	-0.6882	0.0000
-0.8090	-1.1135	-0.8507
-1.3090	-0.4253	-1.3764
-1.3090	0.4253	-0.8507
-0.8090	0.2629	0.0000

-0.5000	-0.6882	0.0000
0.5000	-0.6882	0.0000
0.8090	-1.1135	-0.8507
0.0000	-1.3764	-1.3764
-0.8090	-1.1135	-0.8507
-0.5000	-0.6882	0.0000
0.5000	-0.6882	0.0000
0.8090	0.2629	0.0000
1.3090	0.4253	-0.8507
1.3090	-0.4253	-1.3764
0.8090	-1.1135	-0.8507
0.5000	-0.6882	0.0000
0.8090	0.2629	0.0000

If we take a lower edge of pentagon B such as

0.0000	1.3764	-0.8507
0.8090	1.1135	-1.3764

Then rotate the pentagon through ψ we produce a pentagon on the 3rd layer of our structure.

-0.809017	1.113517	-1.37638
0	1.376382	-0.85065
0.809017	1.113516	-1.376382
0.499999	0.688191	-2.227032
-0.500001	0.688191	-2.227031
-0.809017	1.113517	-1.37638
0	1.376382	-0.85065

Again we can use the symmetry of the Z axis to produce the next 4 faces of the ring by rotating an edited version of this pentagon about the Z axis by 72°, 144°, 216° and 288° to complete the 3rd layer.

-1.309017	0.425325	-0.85065
-0.809017	1.113516	-1.376382
-0.5	0.68819	-2.227032
-0.809017	-0.262866	-2.227031
-1.309018	-0.425325	-1.37638
-1.309017	0.425325	-0.85065
-1.309018	-0.425325	-1.37638
-0.809017	-1.113516	-0.85065
-1.309017	-0.425325	-1.376382
-0.809016	-0.262866	-2.227032

1E-06	-0.850651	-2.227031
0	-1.376382	-1.37638
-0.809017	-1.113516	-0.85065
0	-1.376382	-1.37638
0.809017	-1.113516	-0.85065
0	-1.376382	-1.376382
1E-06	-0.85065	-2.227032
0.809018	-0.262865	-2.227031
1.309017	-0.425326	-1.37638
0.809017	-1.113516	-0.85065
1.309017	-0.425326	-1.37638
1.309017	0.425325	-0.85065
1.309017	-0.425326	-1.376382
0.809017	-0.262865	-2.227032
0.5	0.688192	-2.227031
0.809018	1.113517	-1.37638
1.309017	0.425325	-0.85065
0.809018	1.113517	-1.37638

We have now produced a regular polyhedron hat has symmetry in XY plane with a top face Z=0.0 and a bottom face Z= -2.2270, To centre it about the origin we simply add 2.2270/2 to all the z values.

With a little adjustment we produce a complete regular dodecahedron with unit length edges centred on the origin.

0.8090	0.2629	1.1135
0.0000	0.8507	1.1135
-0.8090	0.2629	1.1135
-0.5000	-0.6882	1.1135
0.5000	-0.6882	1.1135
0.8090	0.2629	1.1135
0.0000	0.8507	1.1135
0.0000	1.3764	0.2629
0.8090	1.1135	-0.2629
1.3090	0.4253	0.2629
0.8090	0.2629	1.1135
0.0000	0.8507	1.1135
-0.8090	0.2629	1.1135
-1.3090	0.4253	0.2629
-0.8090	1.1135	-0.2629
0.0000	1.3764	0.2629
0.0000	0.8507	1.1135

-0.8090	0.2629	1.1135
-0.5000	-0.6882	1.1135
-0.8090	-1.1135	0.2629
-1.3090	-0.4253	-0.2629
-1.3090	0.4253	0.2629
-0.8090	0.2629	1.1135
-0.5000	-0.6882	1.1135
0.5000	-0.6882	1.1135
0.8090	-1.1135	0.2629
0.0000	-1.3764	-0.2629
-0.8090	-1.1135	0.2629
-0.5000	-0.6882	1.1135
0.5000	-0.6882	1.1135
0.8090	0.2629	1.1135
1.3090	0.4253	0.2629
1.3090	-0.4253	-0.2629
0.8090	-1.1135	0.2629
0.5000	-0.6882	1.1135
0.8090	0.2629	1.1135
-0.8090	1.1135	-0.2629
0.0000	1.3764	0.2629
0.8090	1.1135	-0.2629
0.5000	0.6882	-1.1135
-0.5000	0.6882	-1.1135
-0.8090	1.1135	-0.2629
0.0000	1.3764	0.2629
-0.8090	1.1135	-0.2629
-1.3090	0.4253	0.2629
-0.8090	1.1135	-0.2629
-0.5000	0.6882	-1.1135
-0.8090	-0.2629	-1.1135
-1.3090	-0.4253	-0.2629
-1.3090	0.4253	0.2629
-1.3090	-0.4253	-0.2629
-0.8090	-1.1135	0.2629
-1.3090	-0.4253	-0.2629
-0.8090	-0.2629	-1.1135
0.0000	-0.8507	-1.1135
0.0000	-1.3764	-0.2629

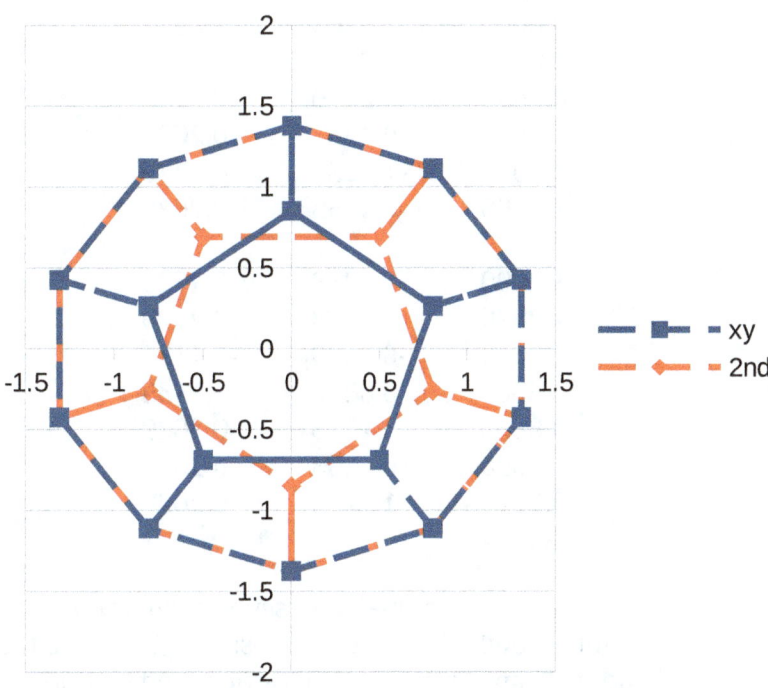

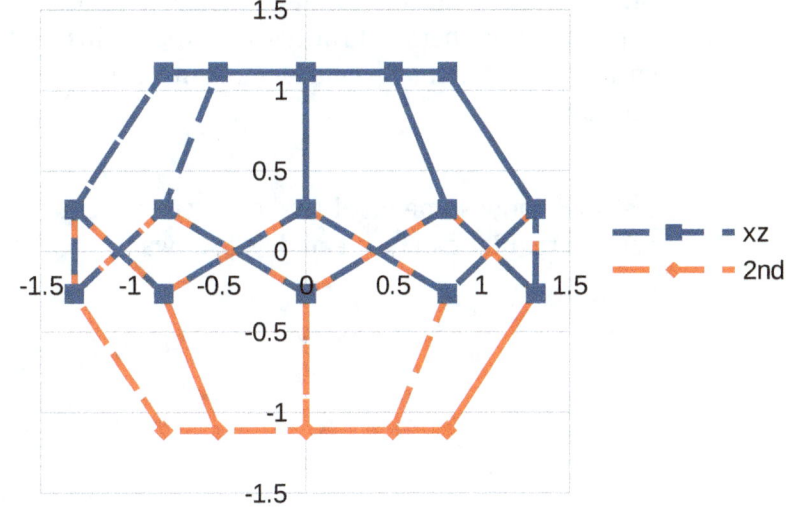

-0.8090	-1.1135	0.2629
0.0000	-1.3764	-0.2629
0.8090	-1.1135	0.2629
0.0000	-1.3764	-0.2629
0.0000	-0.8507	-1.1135
0.8090	-0.2629	-1.1135
1.3090	-0.4253	-0.2629
0.8090	-1.1135	0.2629
1.3090	-0.4253	-0.2629
1.3090	0.4253	0.2629
1.3090	-0.4253	-0.2629
0.8090	-0.2629	-1.1135
0.5000	0.6882	-1.1135
0.8090	1.1135	-0.2629
1.3090	0.4253	0.2629
0.8090	1.1135	-0.2629

There are ways of making pseudo regular polyhedra, they would normally be based on one of these regular polyhedra. The simplest way is to take the mid point of each face, pull it out to the same radius from centre as other vertices and then join it to the vertices of the face to create new edges. This may be done recursively to create ever increasingly more complex structures.

3D regular star structures may be made by taking the mid points of the face of a regular polyhedra and extending them out further than the radius of the regular vertices from the centre of the structure. These may produce amazing structures that are reminiscent of primitive molecules and crystals.

Below I show some that I find of interest. Let us look at the tetragon has 4 faces. The first level pseudo regular can be created as follows.

Rather surprisingly we find we have found a complicated way of creating a cube, however it has served in demonstrating the process.

0.0000	0.5774	-0.2041
-0.5000	-0.2887	-0.2041
0.5000	-0.2887	-0.2041

0.0000	0.5774	-0.2041
-0.5000	-0.2887	-0.2041
0.0000	0.0000	0.6124
0.5000	-0.2887	-0.2041
0.0000	0.0000	0.6124
0.0000	0.5774	-0.2041
0.0000	0.0000	-0.6124
-0.5000	-0.2887	-0.2041
0.0000	0.0000	-0.6124
0.5000	-0.2887	-0.2041
0.0000	-0.5774	0.2041
-0.5000	-0.2887	-0.2041
0.0000	-0.5774	0.2041
0.0000	0.0000	0.6124
0.0000	0.0000	0.6124
0.5000	0.2887	0.2041
0.5000	-0.2887	-0.2041
0.5000	0.2887	0.2041
0.0000	0.5774	-0.2041

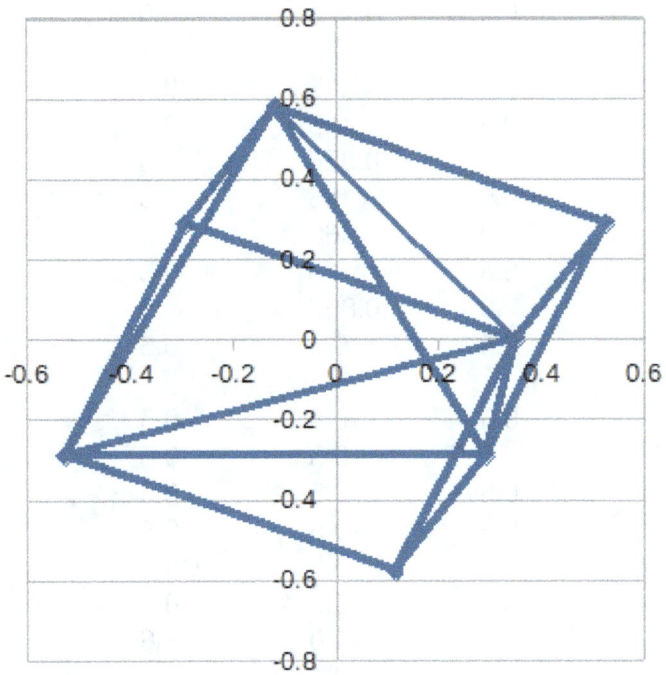

If we do the same with a cube we get much more interesting results.

0.5	0.5	-0.5
0.5	-0.5	-0.5
-0.5	-0.5	-0.5
-0.5	0.5	-0.5
0.0000	0.0000	-0.8660
0.5000	-0.5000	-0.5000
0.0000	0.0000	-0.8660
-0.5000	-0.5000	-0.5000
0.0000	0.0000	-0.8660
0.5	0.5	-0.5
0.5	0.5	0.5
0.5	-0.5	0.5
0.5	-0.5	-0.5
0.5	0.5	-0.5
0.866	0	0
0.5	-0.5	0.5
0.866	0	0
0.5	-0.5	-0.5
0.866	0	0
0.5	0.5	0.5

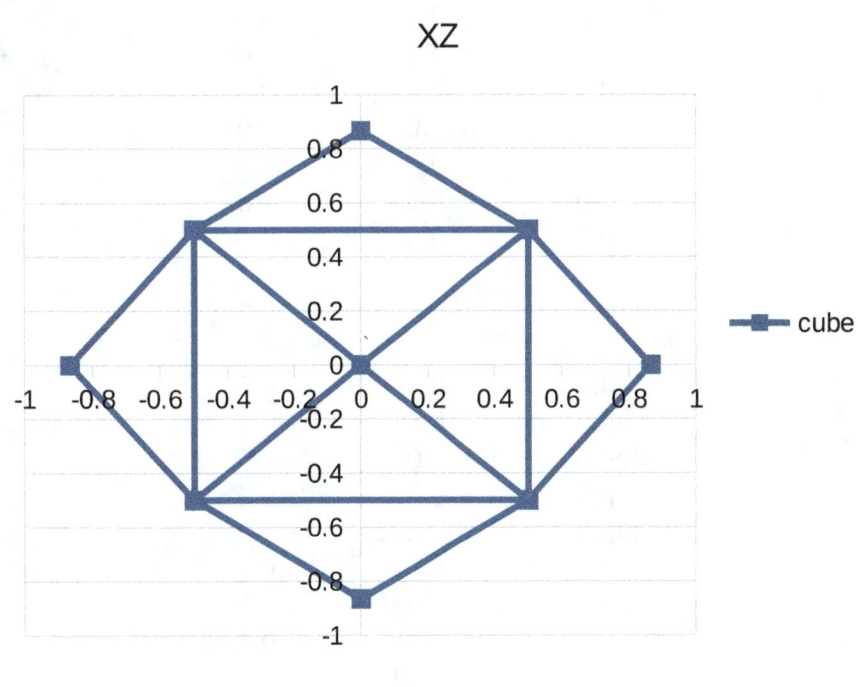

0.5	0.5	0.5
0.5	0.5	-0.5
-0.5	0.5	-0.5
-0.5	0.5	0.5
0	0.866	0
0.5	0.5	-0.5
0	0.866	0
-0.5	0.5	-0.5
0	0.866	0
0.5	0.5	0.5
0.5	-0.5	0.5
0.5	0.5	0.5
-0.5	0.5	0.5
-0.5	-0.5	0.5
0	0	0.866
0.5	0.5	0.5
0	0	0.866
-0.5	0.5	0.5
0	0	0.866
0.5	-0.5	0.5
0.5	-0.5	-0.5
0.5	-0.5	0.5
-0.5	-0.5	0.5
-0.5	-0.5	-0.5
0	-0.866	0
0.5	-0.5	0.5
0	-0.866	0
-0.5	-0.5	0.5
0	-0.866	0
0.5	-0.5	-0.5
-0.5	-0.5	-0.5
-0.5	-0.5	0.5
-0.5	0.5	0.5
-0.866	0	0
-0.5	-0.5	-0.5
-0.866	0	0
-0.5	-0.5	0.5
-0.866	0	0
-0.5	0.5	-0.5

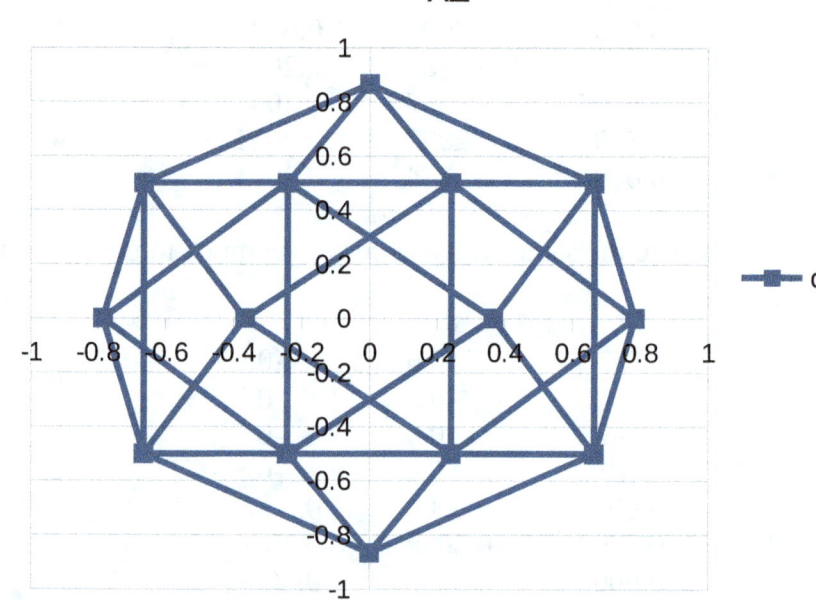

XZ

Cube level 2.

XZ

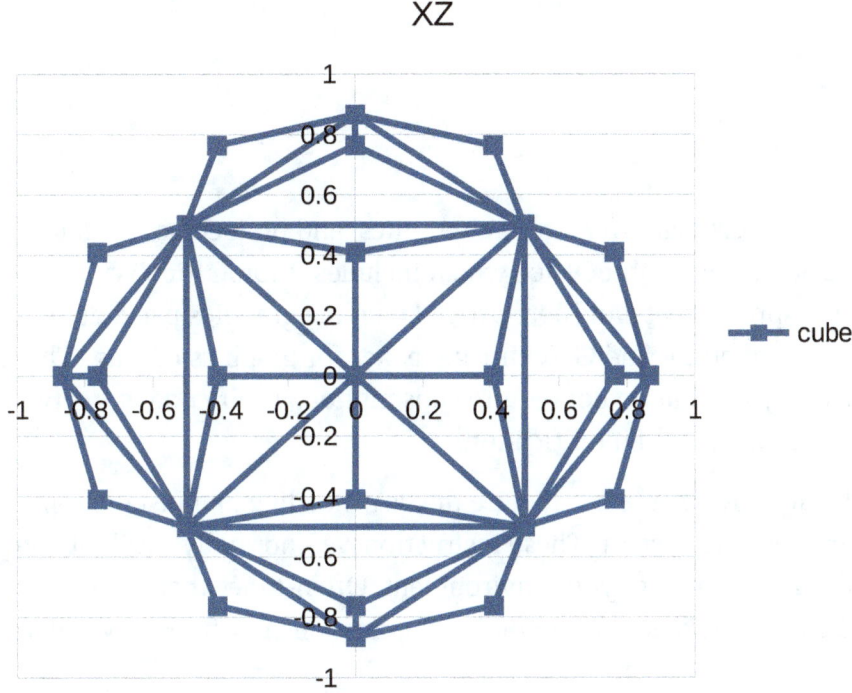

cube

From the octagon we produce the following.

These really come to life when viewed with a 3d visualisation applications such as OpenGL, processing.org (https://processing.org) , 3d viewer (https://3dviewer.net) or any of the other numerous packages that can be used to animate 3 dimensional models. They may demand a degree of modification in format before they will work in the various systems.

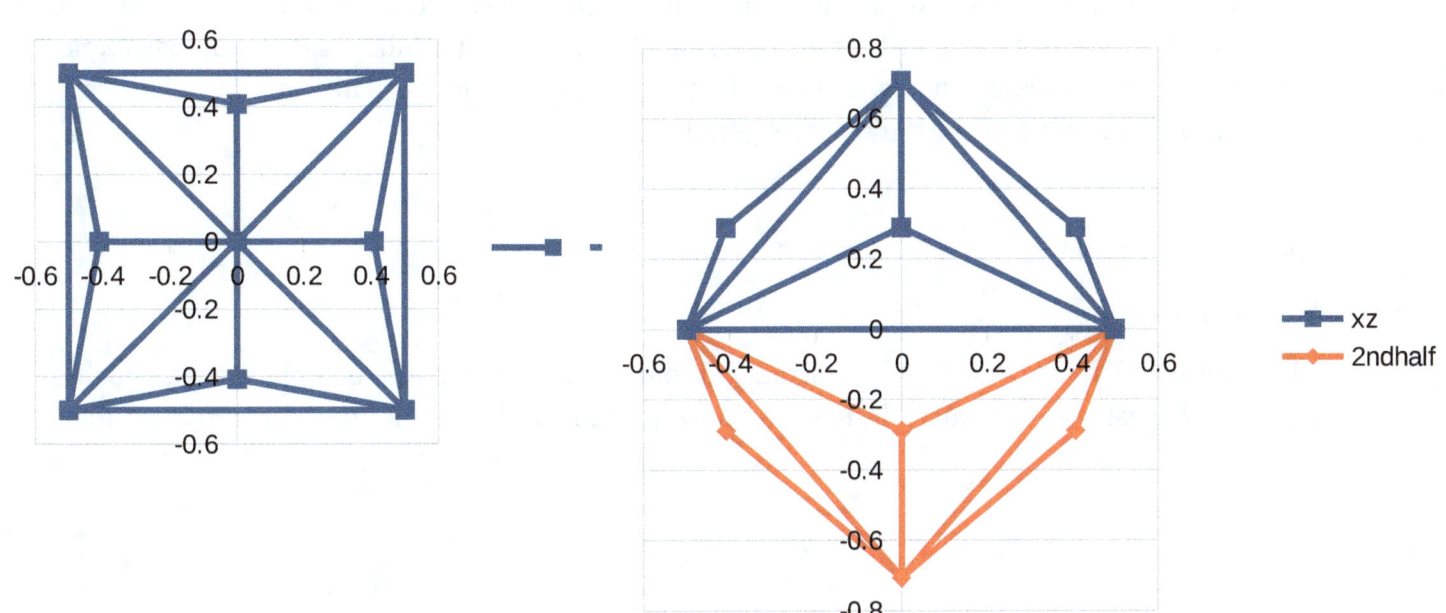

xz
2ndhalf

Appendix

Gnu and Opensource are free to use projects that offer alternative suites and projects to the general public. In this work I used the the Ubuntu suite LibreOffice which includes LibreOffice Writer, the word processor, LibreOffice Calc the spreadsheet and LibreOffice Draw the graphics package, I also used Emacs as a programmable word editor and GNU image maker for graphics editing. These all are royalty free public resourced projects that do not require subscription, Their code is freely available and they promote free development and modification.

Many of the packages in Linux (Ubuntu, Redhat, etc) require the downloading and installation using the relevant installer for your operating system. These instructions are normally available on the internet, but may require tweaking according to your environment. Often-times there is a prerequisite of a pre installation of a base package. Programming applications in Linux may utilise the core gcc compiler.

The Gnu make facility is recommended, it may be used in conjunction with the Windows Cmake facility. Basically when running in terminal these facilities can help compile your c, cpp, Fortran, ADA or other language program. they gather the correct libraries and match the resources your program requires to what you have available in your system.

Processing.org is a java based graphics design package to animate designs and art. A key objective of the foundation is to educate and facilitate the arts community in using electronic platforms to augment their art.

It is ideal to demonstrate the 3 d structures I have designed in the last chapter. When used in a Linux environment it may require an installation of OpenGL and java. After full installation it may require execution from a terminal window. You need to go to the top directory usually .../user/processing4.0b1 then get root status permission with

$sudo -s

and then run it with

$./processing

I have reformatted many of these polyhedra and some of their derivatives into ply format, which is a popular format used in graphic design and I have written readply.pde, a Processing program to

open them and demonstrate them as full 3d images. These are available on my Github
https://github.com/Tunde-Adeyemo/PlatonicSolids

readply.pde

```
/*
readdply.pde allows a local search for a simple ply file
 and displays it in a 3d environment in which we may interactively vary camera angle perspective


 the maximum polygon size is 9 sides
 sept. 2022
*/


import peasy.*;
PeasyCam cam;


float r=0;
float rnc=0.01;
float xp=80;     // xp is a scale factor to be adjusted according to image
int i;

float[] vertx=new float[20000];
int[] tri=new int[1000000];
int nv=0, fcs=0;



void setup() {

  cam = new PeasyCam(this, 100);
  size(800, 800, P3D);

  selectInput("Select a file to process:", "fileSelected");

  noFill();
  stroke(255, 255, 0);
  strokeWeight(2);

  //ortho();
}

void fileSelected(File selection) {
  if (selection == null) {
    println("Window was closed or the user hit cancel.");
  } else {
    println("User selected " + selection.getAbsolutePath());

    BufferedReader reader = createReader(selection);
    String line = null;
    int il=0, ne=0;
```

```
int status=0;
try {
  while ((line = reader.readLine()) != null) {

    if (nv==0)
    {
      if (line.indexOf("vertex")!=-1) {
        String[] sl= splitTokens(line);
        int ix=0;
        while (sl[ix].equals("vertex")==false)
          ix+=1;
        nv=int(sl[ix+1]);
      }
    }
    if (fcs==0)
    {
      if (line.indexOf("face")!=-1) {
        String[] sl= splitTokens(line);
        int ix=0;
        while (sl[ix].equals("face")==false)
          ix+=1;
        fcs=int(sl[ix+1]);
        status =1;
      }
    }
    if (line.indexOf("end_header")!=-1)
    {
      status =2;
      il=0;
      line = reader.readLine();
      // println("status 2");
    }
    if ( status ==1)
      println("vertices ", nv, " ", "faces ", fcs);
    if (status==2)
    {
      line.trim();
      println(line);
      if (line.equals("")== false) {
        String[] sl= splitTokens(line);
        vertx[il*3]=float(sl[0]);
        vertx[il*3+1]=float(sl[1]);
        vertx[il*3+2]=float(sl[2]);
        il++;
        //println(il);
        if (il>nv)
        {
          status =3;
          il=0;
        }
      }
    }

    if (status==3)
    {
      line.trim();
      if (line.equals("")== false) {
```

66

```
        String[] sl= splitTokens(line);
        int sds= int(sl[0]);

        for (i=0; i<=sds; i++) {
          tri[il*10+i]=int(sl[i]);
        }
        il++;
        if (il>fcs)
          status =4;
      }
    }
  }
  reader.close();
  for (i=0; i<nv; i++)
    println(vertx[i*3], " \t ", vertx[i*3+1], " \t ", vertx[i*3+2]);
  for (i=0; i<fcs; i++)
  {
    for (int i2=1; i2<=tri[i*10]; i2++)
      print(tri[i*10+i2], " \t ");
    println("");
  }
}
catch (IOException e) {
  e.printStackTrace();
}
}
}

void draw() {
 if (r> 10.0 ||r<-10.0)
  rnc *= -1;
 int cb=255, cy =0, cr=120;
 //rotateZ(r);
 translate(0, 0, r*-4);
 background(0);
 for (i=0; i<fcs; i++)
 {
  int fc=i%10, fcm= 260/8;
  cb=fc*fcm;
  cy=260-cb;
  cr=cb/2;
  fill(120+cr+cy,cy,cb/2);
  beginShape();
  for (int i2=1; i2<=tri[i*10]; i2++)
    vertex(vertx[tri[i*10+i2]*3] *xp, vertx[tri[i*10+i2]*3+1] *xp, vertx[tri[i*10+i2]*3+2] *xp);

  endShape(CLOSE);
 }

 stroke(255, 0, 0);
 beginShape();   // scale
 vertex(50.0, 0, 0);
 vertex(0.0, 0, 0);
 stroke(0, 255, 0);
 vertex(0.0, 30.0, 0);
 vertex(0.0, 0, 0);
 stroke(0, 0, 255);
```

```
  vertex(0.0, 0.0, 30.0);
  endShape();
  r+=rnc;
}
```

Bibliography

Alan Jeffrey   Mathematics for Engineers and Scientists   Thomas Nelson & Sons ©1969

Jerry Miller   π to 500K Decimal Places   Bandanna Books ©2010

Olatunde Adeyemo   Alien Pi in the Sky? ©2016   www.lulu.com

Olatunde Adeyemo   Matrix Magic ©2010   www.lulu.com

Bash Reference Manual.pdf   Edition 5.1, for Bash Version 5.1.December 2020
Chet Ramey, Case Western Reserve University Brian Fox, Free Software Foundation

Ulla Kirch-Prinz   Peter Prinz   A Complete Guide to Programming in C++
Jones and Bartlett Publishers

Ian Lance Taylor   The GNU configure and build system   Copyright ©1998 Cygnus
Solutions

Richard M. Stallman and the GCC Developer Community   Using the GNU Compiler
Collection   GNU Press Copyright ©1988-2019 Free Software Foundation, Inc.

Torbjörn Granlund and the GMP development team   The GNU Multiple Precision
Arithmetic Library Edition 6.2.1

Makefile - Quick Guide   https://www.tutorialspoint.com/makefile/makefile_quick_guide.htm

https://cmake.org/cmake/help/book/mastering-cmake/index.html

https://cmake.org/documentation/

https://cplusplus.com/reference/

https://www.w3schools.com/

https://www.wolfram.com/mathematica/

https://en.wikipedia.org/wiki/List_of_mathematical_series#Exponential_function

http://www.pi314.net/eng/index.php